Springer Series in Computational Physics

Editors: W. Beiglböck, Heidelberg
H. Cabannes, Paris   S. Orszag, Cambridge, Mass., USA

Frances Bauer
Octavio Betancourt
Paul Garabedian

A Computational Method in Plasma Physics

With 22 Figures

Springer-Verlag
New York   Heidelberg   Berlin

Frances Bauer, Senior Research Scientist
Octavio Betancourt, Research Scientist
Paul Garabedian, Professor of Mathematics, and Director

Courant Mathematics and Computing Laboratory
Courant Institute of Mathematical Sciences
New York University
251 Mercer Street
New York, New York 10012
USA

Editors:

Professor Dr. Wolf Beiglböck

Institut für Angewandte Mathematik
Universität Heidelberg
Im Neuenheimer Feld 5
D-6900 Heidelberg 1
Federal Republic of Germany

Professor Henri Cabannes

Mécanique Théoretique
Université Pierre et Marie Curie
Tour 66, 4, Place Jussieu
F-75005 Paris
France

Professor Stephen A. Orszag

Department of Mathematics
Massachusetts Institute of Technology
Cambridge, Massachusetts 02139
USA

ISBN-13: 978-3-642-85472-9   e-ISBN-13: 978-3-642-85470-5
DOI: 10.1007/978-3-642-85470-5

Library of Congress Cataloging in Publication Data. Bauer, Frances. A computational method in plasma physics. (Springer series in computational physics) Includes bibliographical references. 1. Plasma (Ionized gases)—Data processing. I. Betancourt, Octavio, 1945– joint author. II. Garabedian, Paul R., joint author. III. Title. IV. Series. QC718.B36   530.4′3′0183   78-8982

9 8 7 6 5 4 3 2 1

Preface

In this book, we report on research in methods of computational magneto-hydrodynamics supported by the United States Department of Energy under Contract EY-76-C-02-3077 with New York University. The work has resulted in a computer code for mathematical analysis of the equilibrium and stability of a plasma in three dimensions with toroidal geometry but no symmetry. The code is listed in the final chapter. Versions of it have been used for the design of experiments at the Los Alamos Scientific Laboratory and the Max Planck Institute for Plasma Physics in Garching. We are grateful to Daniel Barnes, Jeremiah Brackbill, Harold Grad, William Grossmann, Abraham Kadish, Peter Lax, Guthrie Miller, Arnulf Schlüter, and Harold Weitzner for many useful discussions of the theory. We are especially indebted to Franz Herrnegger for theoretical and pedagogical comments. Constance Engle has provided outstanding assistance with the typescript. We take pleasure in acknowledging the help of the staff of the Courant Mathematics and Computing Laboratory at New York University. In particular we should like to express our thanks to Max Goldstein, Kevin McAuliffe, Terry Moore, Toshi Nagano and Tsun Tam.

Frances Bauer
Octavio Betancourt
Paul Garabedian

New York
September 1978

Contents

Chapter 1. Introduction   1

1.1 Formulation of the Problem   1
1.2 Discussion of Results   2

Chapter 2. The Variational Principle   4

2.1 The Magnetostatic Equations   4
2.2 Flux Constraints in the Plasma   6
2.3 The Ergodic Constraint   7
2.4 Coordinate System in the Plasma   8
2.5 First Variation of the Potential Energy   10
2.6 Vacuum Region and Force-Free Fields   11
2.7 Variation of the Vacuum Field   12
2.8 Variation of the Free Boundary   14
2.9 Coordinate System in the Vacuum   14
2.10 Accelerated Paths of Steepest Descent   17
2.11 Determination of the Acceleration Coefficients   18

Chapter 3. The Discrete Equations   22

3.1 The Numerical Method   22
3.2 Difference Equations for the Plasma Region   23
3.3 Difference Equations for the Vacuum Region   27
3.4 Iterative Scheme for the Plasma Region   28
3.5 Iterative Scheme for the Vacuum Region   29
3.6 Iterative Scheme for the Free Boundary   30
3.7 Remarks about the Method   31
3.8 Iterative Schemes for Elliptic Equations   33

Chapter 4. Description of the Computer Code   36

4.1 Introduction   36
4.2 Input Data   37

4.3 Printed Output 44
4.4 Glossary 45

Chapter 5. Applications **49**

5.1 Historical Development of the Code 49
5.2 Comparison with Exact Solutions 49
5.3 Unstable High β Stellarator Equilibria 58
5.4 Triangular Cross Sections 63
5.5 High β Tokamaks 67
5.6 Discussion 69

References **70**

Listing of the Code with Comment Cards **72**

1. Output from a Sample Run 72
2. Fortran Listing 78

Index **141**

CHAPTER 1

Introduction

1.1 Formulation of the Problem

In magnetic fusion energy research, a central role is played by toroidal devices for the confinement of a plasma. These devices are essentially of two different types, called the Tokamak and the stellarator. In a Tokamak, which is an axially symmetric configuration having a plane magnetic axis, the toroidal outward drift of the plasma is counterbalanced by the poloidal magnetic field due to a strong toroidal plasma current. In a stellarator, which is a toroidal configuration with a helically deformed magnetic axis, the toroidal drift is offset by a restoring force associated with helical windings, and the net toroidal current is negligible compared to the poloidal current producing the main theta pinch field. Extensive experimental investigations of both types of devices have been conducted. So far the Tokamak work has been more successful and at present dominates the scene.

The partial differential equations of magnetohydrodynamics define a valid isotropic continuum model for mathematical analysis of the toroidal equilibrium of a plasma. When resistivity is neglected, there is a variational principle for the combined magnetic and fluid potential energy that leads to a relatively simple theory of equilibrium and stability [4,22,28]. Even that theory is too complicated, however, to permit exact solutions of many of the problems that arise in the applications. The purpose of this book is to develop a numerical method for the solution of the magnetostatic equations and to present a computer code based on that method for the study of practical questions of equilibrium and stability in plasma physics.

Our intension is to solve problems involving genuinely three-dimensional geometry, such as those associated with the helical windings of a stellarator having no symmetry. Instead of treating the full magnetohydrodynamic equations directly, we calculate equilibria by applying the method of steepest descent to the variational principle for the plasma and vacuum potential energy in a fashion that provides significant information about stability [7,8,12,28]. Therefore, we are able to confine our attention to a reduced system of partial differential equations related to magnetostatics. This simplification does, however, raise some subtle mathematical questions about the formulation of the steady-state problem and the existence of weak solutions [7,20].

Paths of steepest descent are defined by solving an initial value problem for a system of partial differential equations that is expressed in terms of an artificial time parameter. For stable equilibria, the solution approaches a steady state as the artificial time becomes infinite. We introduce an accelerated scheme for which the partial differential equations are of the hyperbolic type. They are more primitive than the full system of magnetohydrodynamic equations, but have many similar properties. In particular, the stability properties of equilibrium solutions are the same.

A computer code has been written to implement our method of finding toroidal equilibrium. Questions of stability can be answered by examining the asymptotic behavior of solutions for large artificial time. A run with adequate resolution can be made in two hours on the CDC 6600 computer. The code is sufficiently fast and accurate to handle three space variables and time with limited computer capacity. For both equilibrium and stability calculations, it is preferable to codes requiring the solution of the full magnetohydrodynamic equations.

1.2 Discussion of Results

The computer code we have developed is most effective for the study of equilibria with medium or high values of the plasma parameter

$$\beta = 2p/(2p + B^2)$$

measuring the ratio of the fluid pressure p to the sum of the fluid pressure and the magnetic pressure $B^2/2$. It is most appropriate for examples where three-dimensional geometry and nonlinear effects play a significant role. Because the code takes into account three space variables as well as the artificial time, there is a severe limitation on how small the mesh sizes can be taken. The resulting truncation errors are not always easy to assess. In general, they take the form of artificial viscosity terms whose effect is in some sense comparable to that of a finite Larmor radius in plasma physics. Both effects tend to reduce growth rates of physically unstable modes.

We have made extensive computer studies of high β stellarators such as the Isar Tl-B at Garching and the Scyllac at the Los Alamos Scientific Laboratory. The calculations enable one to assess the effects of nonlinearity and of a diffuse pressure profile as well as of a vacuum field surrounding the plasma. Unstable equilibria can be determined by examining streak plots of the motion of the plasma corresponding to various helical distortions of the outer conducting coils. Comparable computations have been performed by Barnes and Brackbill [1] at Los Alamos using a three-dimensional code of the Harlow variety for the full magnetohydrodynamic equations. When these computations were used to redesign a set of coils for the final Scyllac experiment, they resulted in a doubling of the containment time, raising it to 50 μsec.

A principal difficulty with high β stellarators has been the instability of the gross $m = 1$, $k = 0$ mode, which shifts the whole plasma to the outer wall. Here m and k indicate the wave numbers in the poloidal and toroidal directions, respectively. Our calculations show that this mode can be stabilized by introducing coils with triangular cross sections [8]. The stabilization depends on the magnetic structure and flux constraints inside the plasma. It enhances the more usual wall stabilization that occurs for low compression ratios. A straight helically symmetric experiment to test this contention is in the construction stage at the Max Planck Institute for Plasma Physics in Garching.

The code is applicable to high β Tokamaks and to Tokamaks with superimposed helical windings. For axially symmetric geometry, it has been used to show that values of β as high as 18 percent can be achieved stable to $m = 1$ by introducing appropriate cross sections. To exhibit the nonlinear and three-dimensional features of the method, we have calculated bifurcated equilibria that are associated with nonlinear saturation of linearly unstable modes.

CHAPTER 2

The Variational Principle

2.1 The Magnetostatic Equations

In magnetohydrodynamics, the equilibrium and stability of a toroidal plasma with density ρ, pressure $p = \rho^\gamma$ and internal energy $e = p/(\gamma - 1)$, confined by a strong magnetic field B, can be analyzed by means of a variational principle [4] for the potential energy

$$E = \iiint \left(\frac{B^2}{2} + \frac{p}{\gamma - 1} \right) dx_1\, dx_2\, dx_3,$$

subject to appropriate constraints. Stationary points correspond to equilibrium solutions; and if the energy has a local minimum, the equilibrium is considered to be stable by definition.

We present a new nonlinear formulation [7] of the standard variational principle of magnetohydrodynamics which is related to that of Kruskal and Kulsrud [28]. Our main objective is to recast the variational principle so that it can more easily be implemented as a computer code. This is to be achieved by using a simple domain for the independent variables in three dimensions, a simple way of introducing constraints, and a minimization procedure that leads to a well-posed problem for a system of partial differential equations involving an artificial time parameter.

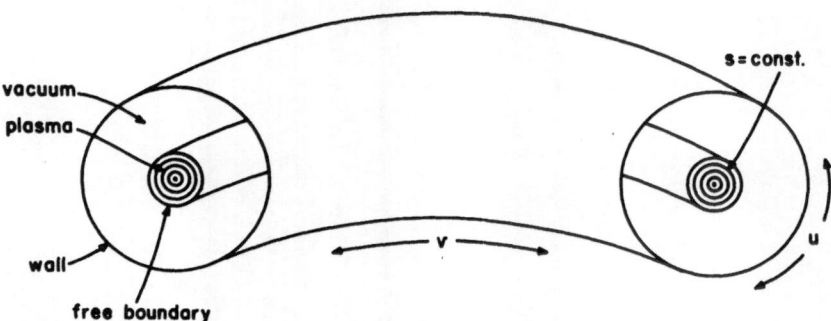

Fig. 2.1   Toroidal geometry.

Let the plasma be contained in a toroidal region Ω_1 of space that is separated by a sharp boundary Γ from an outer vacuum region Ω_2 bounded by a conducting wall C (see Fig. 2.1). We assume that a nested toroidal family of flux surfaces $s = $ const. exists in the plasma region such that $s = 0$ corresponds to the magnetic axis and $s = 1$ corresponds to the free boundary. We denote by u and v variables such as angles with unit periods in the poloidal and toroidal directions, respectively.

Let us minimize the potential energy E subject to the following five constraints:

1. It is required that $\nabla \cdot B = 0$ everywhere.
2. The toroidal and poloidal fluxes within each flux surface in the plasma region are fixed, so that

$$\iint\limits_{s \leq s_0} B \cdot dS = F_T(s_0), \qquad \iint\limits_{s \leq s_0} B \cdot dS = F_P(s_0),$$

where the first integral is evaluated over a disk $v = $ const. and the second integral is evaluated over an annular surface $u = $ const.
3. The mass within each flux tube has a fixed value

$$\iiint\limits_{s \leq s_0} \rho \, dV = M(s_0).$$

4. The total toroidal and poloidal fluxes in the vacuum are fixed. These two conditions can be expressed in the form

$$\iint\limits_{v = \text{const.}} B \cdot dS = F_T^V, \qquad \iint\limits_{u = \text{const.}} B \cdot dS = F_P^V.$$

5. The free surface Γ and the outer wall C are flux surfaces on which the normal component of B vanishes, i.e.,

$$B \cdot v = 0.$$

Note the difference between the flux constraints in the plasma and vacuum regions. In the plasma region the distributions of flux are fixed as functions of s, while in the vacuum region only the two total fluxes are preserved.

The Euler equations for this variational principle are the equations of magnetostatics. In the plasma region we have

$$\nabla p = J \times B, \qquad J = \nabla \times B,$$

where J is the current. On the other hand, in the vacuum region

$$\nabla \times B = 0.$$

The sum $\frac{1}{2}B^2 + p$ of the magnetic and fluid pressures remains continuous across the free boundary Γ. The plasma and vacuum regions must be treated by different methods because the constraints are different in each of them.

The assumption of a nested toroidal family of flux surfaces is justified by the fact that in the time evolution of a magnetohydrodynamic system, the magnetic lines are carried by the fluid and, therefore, the topology is preserved. This leads to a sufficiently simple model, so that the fully three-dimensional problem can be analyzed numerically by solution on a large-scale computer. A more general case allowing for the creation of so-called "islands" has been treated for two-dimensional or axially symmetric geometry by Grad et al. [21].

2.2 Flux Constraints in the Plasma

The first practical difficulty in dealing with the three-dimensional problem is how to prescribe the constraints in a way appropriate for numerical computation. The equation $\nabla \cdot B = 0$ can be integrated by representing the magnetic field B as the cross product [22],

$$B = \nabla s \times \nabla \psi,$$

of the gradients of two scalar flux functions s and ψ. Assuming that the loci $s = $ const. are a nested family of toroidal flux surfaces, we can use s as a Lagrangian coordinate. That is, we switch the role of dependent and independent variables, and s becomes one of our coordinates. Then we prescribe the flux constraints by prescribing the periods of the multiple-valued flux function ψ.

Since B must be single-valued, the most general expression for ψ is

$$\psi = f_1(s)u + f_2(s)v + \lambda(s, u, v),$$

where λ is periodic in u and v. We have for the toroidal flux

$$F_T(s_0) = \iint_{s \leq s_0} B \cdot dS = - \iint_{s \leq s_0} \nabla \times (\psi \nabla s) \cdot dS = \oint \psi \, ds,$$

where the line integral is taken along a curve which bounds the slit disk $v = $ const., $s \leq s_0$. A cut must be introduced along the ray $u = 0$ to make ψ single valued, and the sign is chosen so that the flux is positive in the direction of increasing v. Evaluation of the line integral yields

$$F_T(s_0) = - \int_0^{s_0} f_1(s)ds,$$

where the only nontrivial contribution comes from the cut. Finally, differentiation gives $F'_T(s) = - f_1(s)$.

A similar calculation for the poloidal flux yields $F'_P(s) = f_2(s)$ and, therefore,

$$\psi = -F'_T(s)u + F'_P(s)v + \lambda(s, u, v).$$

The ratio dF_P/dF_T, which we denote by $\mu(s)$, is called the rotational transform.

Clearly, only one function of s need be prescribed to define the fluxes. In the simplest case, where we take s to be the toroidal flux itself so that $F_T(s) = s$, ψ assumes the special form

$$\psi = -u + \mu(s)v + \lambda(s, u, v).$$

However, for practical purposes it is convenient to retain the more general form because it allows us more freedom in the choice of a computational mesh and helps in treating cases where there is a reversal of sign for the main toroidal field.

2.3 The Ergodic Constraint

The relation

$$B \cdot \nabla p = 0$$

implies that the magnetic lines are real characteristics of the magnetostatic equations. In order to formulate a well-posed problem, we want to eliminate these real characteristics from our system of equations. If we assume that the magnetic lines on each flux surface are ergodic, then p must be a function of s alone. It is, therefore, natural to introduce this ergodic constraint on every toroidal flux surface. However, if we wish to arrive at a valid stability analysis we must show that the corresponding relation $\rho = \rho(s)$ yields a minimum of the total internal energy for all choices of $\rho(s, u, v)$ satisfying the basic mass constraint 3 of Section 2.1.

Let

$$D = \frac{\partial(x_1, x_2, x_3)}{\partial(s, u, v)}$$

be the Jacobian of the transformation to the coordinate system s, u, and v. An application of Hölder's inequality asserts that for any fixed s and $\gamma > 1$,

$$M'(s) = m(s) = \iint \rho D \, du \, dv \le \left(\iint \rho^\gamma D \, du \, dv \right)^{1/\gamma} \left(\iint D \, du \, dv \right)^{(\gamma - 1)/\gamma}$$

and equality holds if and only if $\rho = \text{const}$. Hence for fixed $m(s)$ and D, the internal energy

$$E_i = \frac{1}{\gamma - 1} \iiint \rho^\gamma D \, ds \, du \, dv$$

becomes a minimum when

$$\rho = \rho(s) = \frac{m(s)}{\iint D \, du \, dv}.$$

A similar proof of the admissibility of the ergodic constraint can be carried out for $0 < \gamma < 1$, too.

In summary, through the substitutions that have been described we have integrated analytically two of the four magnetostatic equations, while preserving our stability criteria. Moreover, we have eliminated the real characteristics of the system, and that will lead to the formulation of a well-posed problem. The flux and mass constraints have been incorporated explicitly in the formulation, so that an unconstrained minimization problem for the plasma region results.

Observe that the relationship between the magnetic field B and the flux functions s and ψ can be written in the invariant form

$$B_j = \frac{\partial(s, \psi, x_j)}{\partial(x_1, x_2, x_3)}.$$

Thus the expression for the energy E_1 in the plasma region reduces to

$$E_1 = \iiint \frac{D_1^2 + D_2^2 + D_3^2}{2D} \, ds \, du \, dv + \frac{1}{\gamma - 1} \int \frac{m(s)^\gamma \, ds}{(\iint D \, du \, dv)^{\gamma - 1}},$$

where $D_j = \partial(s, \psi, x_j)/\partial(s, u, v)$.

2.4 Coordinate System in the Plasma

We introduce modified cylindrical coordinates r, θ, and z defined by the formulas

$$x_1 = (l + r) \cos \theta,$$
$$x_2 = (l + r) \sin \theta,$$
$$x_3 = z,$$

where l may be interpreted as the large radius of a torus which becomes a cylinder in the limit as $l \to \infty$. Because we have already integrated the equation $B \cdot \nabla p = 0$, we are free to impose on the transformation from the coordinates s, u, and v to r, θ, and z the important restriction

$$\theta = 2\pi v$$

specifying the toroidal angle. Under this hypothesis, the energy E_1 takes the same form as it does in rectangular coordinates, except that we now have

$$D_1 = -\frac{\partial(\psi, r)}{\partial(u, v)}, \qquad D_2 = -L(1 + \varepsilon r)\psi_u, \qquad D_3 = -\frac{\partial(\psi, z)}{\partial(u, v)},$$

$$D = L(1 + \varepsilon r) \frac{\partial(r, z)}{\partial(s, u)},$$

where $L = 2\pi l$ and $\varepsilon = 1/l$. The ratios D_j/D represent the components of the magnetic field B in the r, θ, and z directions, respectively. We can set $\varepsilon = 0$, keeping L fixed, to obtain a cylinder of length L with periodic boundary conditions. Note also that with the special prescription for v only two-dimensional Jacobians are needed.

We could consider ψ as a Lagrangian coordinate, i.e., as a fixed function of s, u, and v, and minimize E_1 over all periodic mappings $r(s, u, v)$, $z(s, u, v)$ of the cube

$$\Omega: 0 \leq s \leq 1; \quad 0 \leq u \leq 1; \quad 0 \leq v \leq 1$$

onto a specified plasma region. Variation of the independent variables r and z and the fact that p is a function of s alone would show that the Euler equations of the new extremal problem reduce to the magnetostatic equations. However, there are several difficulties with such a formulation. The solution for ψ is not unique, since we can add any function of s alone to ψ without changing the values of B. This is reflected in the fact that the solution for the mapping is not unique, and corresponding compatibility conditions due to the toroidal geometry must be satisfied. Moreover, the magnetic axis is a singular curve in our coordinate system, which makes it difficult to write equations for r and z at $s = 0$. Finally, the boundary condition for r and z at $s = 1$ is nonlinear.

It is more effective to replace the physical coordinates r and z as dependent variables by a combination of the flux function $\psi = \psi(s, u, v)$ and a dimensionless radius $R = R(s, u, v)$ related to r and z by the formulas

$$r = r_0(v) + R(s, u, v)[r_1(u, v) - r_0(v)],$$

$$z = z_0(v) + R(s, u, v)[z_1(u, v) - z_0(v)],$$

where $r = r_0(v)$, $z = z_0(v)$ are the equations of the magnetic axis and $r = r_1(u, v)$, $z = z_1(u, v)$ are the equations of the free boundary Γ. The function R serves to define the geometry of the flux surfaces $s = $ const. The boundary conditions on R require that it be periodic in u and v and that $R = 0$ at $s = 0$, and $R = 1$ at $s = 1$. There are no boundary conditions on ψ other than the poloidal and toroidal periodicity requirements already indicated. The functions r_0 and z_0 must be periodic and they must be found as part of the answer to the minimum problem. In terms of R, the Jacobian D reduces to the simple expression

$$D = LH(1 + \varepsilon r)RR_s,$$

where

$$H(u, v) = (z_1 - z_0)\frac{\partial r_1}{\partial u} - (r_1 - r_0)\frac{\partial z_1}{\partial u}.$$

2.5 First Variation of the Potential Energy

Making perturbations $\delta\psi$, δR, δr_0, and δz_0 of the dependent variables ψ, R, r_0, and z_0, we obtain, after integration by parts, an expression of the form

$$\delta E_1 = -\iiint (L_1(\psi)\delta\psi + L_2(R)\delta R)ds\,du\,dv$$

$$- \int (L_3(r_0)\delta r_0 + L_4(z_0)\delta z_0)dv$$

for the first variation of the energy E_1 in the plasma region. A calculation shows that the operators $L_1(\psi)$, $L_2(R)$, $L_3(r_0)$ and $L_4(z_0)$ occurring here are defined by the relations

$$L_1(\psi) = \frac{\partial}{\partial u}\frac{[r_v^2 + L^2K^2 + z_v^2]\psi_u - [r_u r_v + z_u z_v]\psi_v}{D}$$

$$+ \frac{\partial}{\partial v}\frac{[r_u^2 + z_u^2]\psi_v - [r_u r_v + z_u z_v]\psi_u}{D},$$

$$L_2(R) = (r_1 - r_0)\left\{\frac{\partial}{\partial u}\frac{\psi_v[\psi_v r_u - \psi_u r_v]}{D} + \frac{\partial}{\partial v}\frac{\psi_u[\psi_u r_v - \psi_v r_u]}{D}\right.$$

$$+ L\varepsilon\left(\frac{DP}{LK} - \frac{LK\psi_u^2}{D}\right)\right\}$$

$$+ (z_1 - z_0)\left\{\frac{\partial}{\partial u}\frac{\psi_v[\psi_v z_u - \psi_u z_v]}{D} + \frac{\partial}{\partial v}\frac{\psi_u[\psi_u z_v - \psi_v z_u]}{D}\right\}$$

$$- LHR\frac{\partial}{\partial s}(PK),$$

$$L_3(r_0) = \iint\left[(1 - R)\left\{\frac{\partial}{\partial u}\frac{\psi_v[\psi_v r_u - \psi_u r_v]}{D} + \frac{\partial}{\partial v}\frac{\psi_u[\psi_u r_v - \psi_v r_u]}{D}\right.\right.$$

$$\left.\left.+ L\varepsilon\left(\frac{DP}{LK} - \frac{LK\psi_u^2}{D}\right)\right\} - PLKRR_s\frac{\partial z_1}{\partial u}\right]ds\,du,$$

$$L_4(z_0) = \iint\left[(1 - R)\left\{\frac{\partial}{\partial u}\frac{\psi_v[\psi_v z_u - \psi_u z_v]}{D} + \frac{\partial}{\partial v}\frac{\psi_u[\psi_u z_v - \psi_v z_u]}{D}\right\}\right.$$

$$\left.+ PLKRR_s\frac{\partial r_1}{\partial u}\right]ds\,du,$$

where $P = \frac{1}{2}B^2 + p$ and $K = 1 + \varepsilon r$.

The Euler equations $L_1(\psi) = 0$ and $L_2(R) = 0$, asserting that E_1 is a stationary functional of ψ and R, imply that for magnetostatics,

$$\nabla s \cdot J = 0, \qquad \nabla\psi \cdot J = p'(s).$$

The first of these can be viewed as an elliptic equation for ψ in its dependence on the two variables u and v within the flux surfaces $s = $ const. Similarly, the second is an elliptic equation for s embedded in the two-dimensional flux surfaces $\psi = $ const. These two equations, together with the ergodic constraint $p = p(s)$, imply $\nabla p = J \times B$. They are of nonstandard type in three-dimensional space.

The corresponding Euler equations $L_3(r_0) = 0$ and $L_4(z_0) = 0$ for r_0 and z_0 can be written as

$$L_3(r_0) = \iint (1 - R)(\nabla p - J \times B) \cdot \hat{e}_r D \, ds \, du = 0$$

and

$$L_4(z_0) = \iint (1 - R)(\nabla p - J \times B) \cdot \hat{e}_z D \, ds \, du = 0,$$

where $\hat{e}_r$ and $\hat{e}_z$ are unit vectors in the r and z directions. They represent a weighted average of the magnetostatic forces on the cross section $v = $ const.

For the case in which there is no vacuum region, the functions r_1 and z_1 represent the equations of the outer conducting wall C, and the problem reduces to minimizing the expression for the potential energy E_1 in the plasma region alone as a functional of ψ, R, r_0, and z_0. When a vacuum region is present, the location of the free boundary Γ must be found by considering the contribution of the vacuum region to the potential energy. This is the subject of the next section.

2.6 Vacuum Region and Force-Free Fields

Consider the total potential energy, which we write as

$$E = E_1 + E_2 = \iiint_{\Omega_1} \left(\frac{1}{2} B^2 + \frac{p}{\gamma - 1} \right) dV + \iiint_{\Omega_2} \frac{1}{2} B^2 \, dV,$$

where dV is the volume element and Ω_1 and Ω_2 are the plasma and vacuum regions, respectively (see Fig. 2.1). The variation of E with respect to the magnetic field B and pressure p in the plasma region has already been discussed. Now we consider the problem of minimizing E with respect to variations of the vacuum magnetic field B and the free boundary Γ subject to the vacuum flux constraints 4 and 5 of Section 2.1. Again we emphasize that in this case it is not the flux distribution $\mu(s)$ but just the total poloidal and toroidal fluxes in the vacuum region that are fixed.

At first glance, one might think that we could proceed in the same manner as before and express the vacuum magnetic field as $B = \nabla s \times \nabla \psi$ to satisfy $\nabla \cdot B = 0$. However, there are several reasons why that is not the right way to proceed.

First of all, the flux surfaces in the vacuum region will not, in general, be nested tori for genuinely three-dimensional geometry [32]. While in the plasma region the topology of the magnetic lines has an invariant structure, no such requirement holds for the vacuum magnetic field, and topological assumptions are not natural.

Second, even if we were to accept the constraint of nested toroidal surfaces, we would have to minimize the energy E_2 with respect to the function

$$F(s) = \int_{s_0}^{s} \mu(\sigma)d\sigma$$

over the vacuum region. Making a variation $\delta\mu$ shows that the corresponding Euler equation has the form

$$I(s) = \oint B \cdot dx = \text{const.}$$

where I is the net current through the toroidal flux tube $s = $ const. It is only when sufficient regularity is assumed that this equation, together with the force-free field condition $J \times B = 0$, implies $J = 0$. In more general cases, F might be continuous but not differentiable and the answer to the minimum problem might be a weak solution with surface current sheets through the points where μ is discontinuous. On such surfaces, the magnetic lines need not be ergodic and there could be surface currents with alternating signs.

Because of these considerations, we introduce a reciprocal variational problem [5,6]. It is well known that if we minimize E_2 with respect to B subject to $\nabla \cdot B = 0$, the corresponding Euler equation is $\nabla \times B = 0$. We now consider the reciprocal problem, which is to find the stationary point for E_2 subject to $\nabla \times B = 0$ and subject to a suitable formulation of the flux constraints 4. The corresponding Euler equation is $\nabla \cdot B = 0$, and both problems have the same solution. However, in the reciprocal case the vacuum energy is a maximum with respect to variations of B, as we shall prove in the next section. The advantage of the constraint $\nabla \times B = 0$ is that it is easily imposed by means of a scalar potential, but its disadvantage is that we have to deal with a minimax problem rather than a straight minimum problem for E. In principle, one should determine the vacuum field for each position of the free boundary Γ and then minimize with respect to the other dependent variables. In practice, though, it suffices to iterate the vacuum field equation more often than the rest of the dependent variables.

2.7 Variation of the Vacuum Field

We integrate the constraint $\nabla \times B = 0$ by setting

$$B = \nabla\phi = c_1\nabla\phi_1 + c_2\nabla\phi_2,$$

where the potentials ϕ_1 and ϕ_2 are associated with unit currents $\int d\phi_j$ in the poloidal and toroidal directions, respectively, but have zero periods in the conjugate directions. The Euler equation stating that ϕ is a stationary point of E_2 implies that $\Delta\phi_i = 0$ in Ω_2, and that $\partial\phi_i/\partial\nu = 0$ on Γ and C. Note that we have here a natural boundary condition in the sense of the calculus of variations, which means that the boundary condition is a consequence of the extremal property.

Since it is the fluxes F_P^V and F_T^V that must be fixed, we require

$$a_{11}c_1 + a_{12}c_2 = F_P^V,$$

$$a_{21}c_1 + a_{22}c_2 = F_T^V,$$

where the matrix A with elements

$$a_{ij} = \iiint_{\Omega_2} \nabla\phi_i \cdot \nabla\phi_j \, dV, \qquad i, j = 1, 2,$$

is called the inductance matrix. Let c denote the vector with components c_1 and c_2, and f the vector with components F_P^V and F_T^V. We have

$$E_2 = \tfrac{1}{2} \iiint_{\Omega_2} (\nabla\phi)^2 \, dV = \tfrac{1}{2} c' A c,$$

where c' is the transpose of c. The matrix A is symmetric and positive definite, and using the flux constraints we obtain

$$E_2 = \tfrac{1}{2} f' A^{-1} f.$$

We intend to prove that E_2 is a maximum with respect to variations of ϕ for $Ac = f$ fixed. Let A_0 be the matrix corresponding to the stationary solutions with $\Delta\phi_i = 0$, $\partial\phi_i/\partial\nu = 0$. Since A and A_0 are symmetric and positive definite, we can find an orthogonal basis such that

$$c' A_0 c = \lambda_1 \xi_1^2 + \lambda_2 \xi_2^2, \qquad c' A c = \xi_1^2 + \xi_2^2,$$

where ξ_1 and ξ_2 are the components of c in the new basis, and λ_1 and λ_2 are the roots of the equation $|A_0 - \lambda A| = 0$. Correspondingly, we can write

$$f' A_0^{-1} f = \mu_1 \eta_1^2 + \mu_2 \eta_2^2, \qquad f' A^{-1} f = \eta_1^2 + \eta_2^2,$$

where μ_1 and μ_2 are the roots of $|A_0^{-1} - \mu A^{-1}| = 0$. This implies that $\mu_i = 1/\lambda_i$. Now by Dirichlet's principle we have

$$c' A_0 c \le c' A c$$

for all c and, therefore, $\lambda_i \le 1$. As a consequence, $\mu_i \ge 1$, so that

$$f' A_0^{-1} f \ge f' A^{-1} f$$

for any ϕ_i, which completes the proof.

2.8 Variation of the Free Boundary

Next we consider the variation of the energy E as a functional of the position of the free boundary Γ. Let δv be an arbitrary perturbation of Γ along its outer normal. A direct calculation shows that [6]

$$\delta E_2 = -\tfrac{1}{2}c'\delta Ac,$$

and Hadamard's variational formula for harmonic functions [17] implies that

$$\delta a_{ij} = - \iint_{\Gamma} \nabla \phi_i \cdot \nabla \phi_j \delta v \, dS,$$

where dS is the surface area element on Γ. A similar calculation for the plasma region shows that

$$\delta E_1 = - \iint_{\Gamma} (\tfrac{1}{2}B_1^2 + p)\delta v \, dS,$$

which, together with the above equations, leads to

$$\delta E = - \iint_{\Gamma} (\tfrac{1}{2}B_1^2 + p - \tfrac{1}{2}B_2^2)\delta v \, dS,$$

where B_1 and B_2 stand for the limiting values of the magnetic field coming from the plasma and the vacuum, respectively. This means that $\tfrac{1}{2}B^2 + p$ must be continuous across Γ for a solution of the variational problem.

2.9 Coordinate System in the Vacuum

The basic difficulty in developing a numerical scheme for a free boundary problem is to handle the changing shape and location of the free boundary. If the problem involves only two independent variables, conformal mapping techniques can be used to solve it in a fixed auxiliary domain [10]. Then the region of the solution is determined as the conformal image of that domain.

In the general case of three independent variables no such conformal mapping exists, but the basic idea of mapping the physical region onto a fixed auxiliary domain Ω can still be used. Since the mapping is not conformal, the resulting equations in Ω will be more complicated, which in turn means that finding the solution will require a greater amount of computation. However, the advantage of having a fixed domain in which to solve difference equations far outweighs the disadvantage of the extra computation. This also provides an ideal framework in which to use fast, direct methods for solving the resulting difference equations. We shall have more to say about that later.

We choose to formulate the vacuum energy problem as a minimum problem for Dirichlet's integral $\iiint (\nabla \phi)^2 \, dV$ rather than to start from Laplace's equation. This approach has several advantages, one of them being that the boundary condition $\partial \phi / \partial v = 0$ is a consequence of the minimization and need not be imposed as a special requirement. Another advantage is that the resulting Euler equation will be in conservation form and, therefore, the compatibility condition for a solution of the Neumann problem will be satisfied. These properties can easily be extended to difference equations by using a discrete variational principle in a fashion suggested by the finite element method.

We start with the cylindrical coordinate system of Section 2.4 and put $\theta = 2\pi v$ again. Consider the mapping of the cube

$$\Omega : 0 \le s \le 1; \qquad 0 \le u \le 1; \qquad 0 \le v \le 1$$

onto the vacuum region Ω_2 given by

$$r = r_1(u, v) + s[r_2(u, v) - r_1(u, v)],$$

$$z = z_1(u, v) + s[z_2(u, v) - z_1(u, v)],$$

where $r = r_2(u, v)$ and $z = z_2(u, v)$ are the equations of the outer conducting wall C and $r = r_1(u, v)$, $z = z_1(u, v)$ are, as before, the equations of the free boundary Γ. The Dirichlet integral can be written as

$$\iiint_{\Omega_2} (\nabla \phi)^2 \, dV = \iiint (a\phi_s^2 + b\phi_u^2 + c\phi_v^2 + 2d\phi_s\phi_u$$

$$+ 2e\phi_s\phi_v + 2f\phi_u\phi_v)ds \, du \, dv,$$

where

$$L = \frac{2\pi}{\varepsilon}, \qquad K = 1 + \varepsilon r, \qquad \Delta = \frac{\partial(r, z)}{\partial(s, u)},$$

and

$$a = \frac{LK(r_u^2 + z_u^2 + e^2)}{\Delta}, \qquad b = \frac{LK(r_s^2 + z_s^2 + f^2)}{\Delta},$$

$$c = \frac{\Delta}{LK}, \qquad d = \frac{LK(ef - r_u r_s - z_u z_s)}{\Delta}$$

$$e = \frac{r_u z_v - r_v z_u}{LK}, \qquad f = \frac{r_v z_s - r_s z_v}{LK}.$$

The periodicity conditions on ϕ_1 and ϕ_2 now become

$$\phi_i(s, u + 1, v) = \phi_i(s, u, v) + \delta_{i1},$$

$$\phi_i(s, u, v + 1) = \phi_i(s, u, v) + \delta_{i2}$$

for $i = 1, 2$, where δ_{ij} is the Kronecker delta. In the new coordinates s, u, v over Ω, the Euler equation, equivalent to Laplace's equation, appears in the conservation form

$$\frac{\partial}{\partial s}(a\phi_s + d\phi_u + e\phi_v) + \frac{\partial}{\partial u}(b\phi_u + d\phi_s + f\phi_v)$$

$$+ \frac{\partial}{\partial v}(c\phi_v + e\phi_s + f\phi_u) = 0,$$

with boundary conditions at $s = 0, 1$ given by

$$a\phi_s + d\phi_u + e\phi_v = 0.$$

To specify the free boundary variation, we write the equations of Γ in terms of a dimensionless radius $g = g(u, v)$ as

$$r_1(u, v) = r_3(v) + g(u, v)[r_2(u, v) - r_3(v)],$$

$$z_1(u, v) = z_3(v) + g(u, v)[z_2(u, v) - z_3(v)],$$

where $r_3(v)$, $z_3(v)$ are the equations of a curve defining a new origin of co-ordinates in each meridian plane $v = $ const. This closed curve can be chosen to follow the shape of the outer wall, so that g becomes a slowly varying function of u and v.

Making a perturbation δg, we obtain after integration by parts the variation of the energy due to a shift of the free boundary in the form

$$\delta E = - \iint M(g)\delta g \, du \, dv.$$

Here $M(g)$ is defined, following Section 2.8, by the formula

$$M(g) = LK[(r_2 - r_3)z_u - (z_2 - z_3)r_u][\tfrac{1}{2}B_1^2 + p - \tfrac{1}{2}B_2^2].$$

From Section 2.4 we obtain the expression for the plasma magnetic pressure

$$B_1^2 = \left(\frac{[r_u^2 + z_u^2]\psi_v^2 + [(LK)^2 + r_v^2 + z_v^2]\psi_u^2 - 2[r_u r_v + z_u z_v]\psi_u \psi_v}{(LK)^2(HRR_s)^2}\right).$$

On the other hand, using the boundary conditions for ϕ_i on Γ, we have for the vacuum magnetic pressure there

$$B_2^2 = \left(\frac{[r_u^2 + z_u^2]\phi_v^2 + [(LK)^2 + r_v^2 + z_v^2]\phi_u^2 - 2[r_u r_v + z_u z_v]\phi_u \phi_v}{([LK]^2[r_u^2 + z_u^2] + [r_v z_u - r_u z_v]^2)}\right).$$

The free boundary condition is, of course, just the first-order partial differential equation $M(g) = 0$ for g.

This completes our formulation of the variational principle. The formulas that we have derived will be the main tool to set up a minimization procedure to be described in the next section.

2.10 Accelerated Paths of Steepest Descent

We propose to solve the magnetostatic boundary value problem for ψ, R, r_0, z_0, and g by considering paths of steepest descent associated with the minimum energy principle for E. We assume that for any g we have solved the vacuum equations for the potentials ϕ_i exactly, so that the vacuum energy is a functional of g alone.

Letting the unknown functions depend on an artificial time parameter t, we define an accelerated path of steepest descent by means of the system of partial differential equations

$$a_1\psi_{tt} + e_1\psi_t = L_1(\psi),$$

$$a_2 R_{tt} + e_2 R_t = L_2(R),$$

$$a_3(r_0)_{tt} + e_3(r_0)_t = L_3(r_0),$$

$$a_3(z_0)_{tt} + e_3(z_0)_t = L_4(z_0),$$

$$e_4 g_t = M(g),$$

where the operators on the right come from the Euler equations found in Sections 2.5 and 2.9. The form of the equations is motivated by the method of steepest descent, the conjugate gradient method, and the second-order Richardson method. The coefficients a_j are to be determined so that the artificially time-dependent system becomes hyperbolic, while the e_j are supposed to be large enough to maintain descent.

The first thing to be noticed is that the system is chosen so that the energy E becomes a decreasing function of t. If $a_j = 0$, we have the method of steepest descent and E_t is negative, as can be seen from our formulas for the first variation. Furthermore, the path is chosen in the direction of maximum descent. However, for $a_j = 0$ the system is not adequate because the type of the differential operators on the right, which, with the exception of $M(g)$, are second order in the space variables, is nonstandard. Thus we have added second-order time derivatives so as to obtain a hyperbolic system. The convergence to a steady-state solution would be prohibitively slow without such acceleration terms. This explains our use of the second-order Richardson method, which is more or less equivalent to the conjugate gradient method in the present case.

The idea is to compute solutions of the artificially time-dependent system in the limit as $t \to \infty$. If the associated plasma equilibrium is stable, the energy has a relative minimum and the answer will converge to a steady-state solution of the magnetostatic equations. If, on the other hand, the equilibrium is unstable, the energy has a saddle point, and the artificially time-dependent solution will diverge from equilibrium following essentially the most unstable eigenfunction. If only quadratic terms are kept in an expansion of E about equilibrium, this procedure reduces to the standard variational principle of magnetohydrodynamics [4].

By choosing the coefficients a_j and e_j appropriately, we are able to study questions of both equilibrium and stability with far less computational effort than would be necessary if we examined dependence on the physical time instead (cf. [11]). In some sense, our artificially time-dependent system of partial differential equations may be interpreted as a primitive model of magnetohydrodynamics. A similar approach has been proposed by Chodura and Schlüter [12].

Since our formulation is nonlinear, we can study problems that are beyond the scope of the usual linearized stability analysis. For example, we can consider the case of a solution which is linearly stable but becomes unstable under large perturbations. Conversely, we can investigate the problem of bifurcation by perturbing an unstable equilibrium and seeing if the result converges to a different stable solution. This is sometimes referred to as saturation.

The same method could be used to do linearized stability analysis for problems with axial or helical symmetry. If we impose the symmetry condition on the formula for the energy, then the variational principle leads to a two-dimensional problem for the equilibrium solution. One equation can be integrated explicitly, and we are led to a single equation for a scalar potential. If we linearize about this solution, a Fourier analysis can be done with respect to the ignorable coordinate. Here our method has the advantage that the coordinate system used for the equilibrium problem is also convenient for solving the stability problem (cf. [23]). We hope to work on implementing this approach in a future publication.

A major contribution is the great generality allowed by our coordinate system. Since this follows the motion of the plasma, we can compute solutions with large deviations from axial symmetry but still use unknown functions that are slowly varying. Thus it is relatively easy to study the effect of wall shape or compression ratio on the stability properties of the solution.

2.11 Determination of the Acceleration Coefficients

The role of the coefficients a_j is to bring the system of partial differential equations of Section 2.10 into the hyperbolic type and provide it with appropriate characteristics. The a_j will later be selected to meet the Courant–

Friedrichs–Lewy criterion for stability of analogous difference equations on a given mesh (cf. [17]). Assuming them to be fixed, the rate of convergence in t is governed by the first-order coefficients e_j. To look for the best device to accelerate the convergence, let us consider the example of a simple scalar equation. It will be obvious how our conclusions are to be generalized to handle the plasma physics problem.

Suppose that $L(\psi)$ is a linear second-order differential operator in the space variables and that we are minimizing a functional E whose first variation is

$$\delta E = - \iiint L(\psi)\delta\psi \, dV.$$

The associated paths of steepest descent are defined by

$$a\psi_{tt} + e\psi_t = L(\psi).$$

Let Ψ be an eigenfunction corresponding to a negative eigenvalue $-\omega^2$ of L and set $\psi = e^{\lambda t}\Psi$, where λ satisfies the dispersion relation

$$a\lambda^2 + e\lambda = -\omega^2.$$

In order to maintain descent, we need

$$E_t = - \iiint (a\lambda + e)\psi_t^2 \, dV < 0,$$

which requires $e/a > |\lambda|$. If we choose e independent of t, we are forced to have e/a greater than the largest value of $|\lambda|$ that occurs in a given distribution of initial data. For $e \gg a$, we have the asymptotic relation

$$\lambda \approx - \frac{\omega^2}{e}.$$

If ω is small, which is the case we are primarily interested in, the resulting convergence rate is much too slow. This can be interpreted to mean that the artificial time t scales like the square root of real time. Similar considerations apply to any positive eigenvalue of L.

To accelerate the method, we choose e to be proportional to the dominant growth rate λ, which may be either positive or negative and may vary with time. Then, according to the dispersion relation, we obtain

$$\lambda \approx \omega \text{ const.},$$

so the artificial time scales like real time rather than like its square root.

To implement this idea, consider the least-squares error

$$F(t) = \iiint L(\psi)^2 \, dV.$$

In terms of the eigenfunctions Ψ_n of L, this can be expanded as

$$F(t) = \sum A_n \exp\{2\lambda_n t\} \iiint L(\Psi_n)^2 \, dV,$$

where for the sake of simplicity we have assumed orthogonality. The values of $|F_t/F|$ averaged over a number of time cycles provide a good measure of the dominant growth rate λ. By setting

$$e(t) = \tau a \left| \frac{F_t}{F} \right|$$

for a suitable value of the constant $\tau \geq 1$, the best rate of convergence is achieved and simultaneously an estimate is obtained of the growth rate ω of the least favorable mode for an unstable equilibrium. To ascribe a physical meaning to this growth rate in practice, however, comparison must be made with some example in which an Alfven transit time is known from other considerations.

The procedure we have described for acceleration by means of a variable convergence factor $e = e(t)$ significantly enhances the method of steepest descent, which is prohibitively slow in its usual formulation. The same procedure is applicable to the problem of estimating the relaxation factor for the method of successive over-relaxation in a more general context. This will be described in the next chapter in connection with the solution of Laplace's equation for the vacuum region. In practice no extra computational work is required, since the operator $L(\psi)$ must be computed in any case. Rates of convergence can be improved by as much as a factor of ten in typical cases.

For the partial differential equation of the free surface, an exception has to be made because it is only of the first order. However, no acceleration is called for in that case anyway, so the coefficient of the time derivative can be assigned in a more obvious fashion. It then turns out that the previous assertions about growth rates remain valid even with a free surface included in the model. However, for the free surface model the convergence of the solution is markedly improved if we allow the origin of the coordinate system for the free boundary to move with the plasma. This is accomplished by writing differential equations defining paths of steepest descent for $r_3(v)$ and $z_3(v)$ which only involve averages with respect to u of the free boundary equation. These are given by

$$e_5(r_3)_t = \int (1 - g)K[\tfrac{1}{2}B_1^2 + p - \tfrac{1}{2}B_2^2]z_u \, du,$$

$$e_5(z_3)_t = - \int (1 - g)K[\tfrac{1}{2}B_1^2 + p - \tfrac{1}{2}B_2^2]r_u \, du.$$

Such a rezoning minimizes mesh distortion because the origin of the coordinate system follows the plasma shape. For example, helical excursion or translation of the plasma column is described primarily by the moving origin itself rather than by large distortions of the function g. Thus truncation errors are minimized, too.

CHAPTER 3

The Discrete Equations

3.1 The Numerical Method

In order to develop a scheme for the numerical computation of magneto-hydrodynamic equilibria, we apply the ideas used in the previous chapter to formulate a discrete variational principle for the potential energy E. This is accomplished by writing down a second-order accurate discrete approximation of E based on a rectangular mesh. To arrive at finite difference equations for equilibrium, we set the derivatives of the approximation of E with respect to the nodal values of the unknowns equal to zero.

There are several advantages to this approach. The first is that the resulting set of difference equations satisfies compatibility conditions for the existence of solutions to the equations for ψ and ϕ arising from the toroidal geometry. In neither case is the solution unique. To any solution of the ψ equation one can add an arbitrary function of s, and to any solution of the ϕ equation one can add an arbitrary constant and still have a solution. This means that the matrix of coefficients of the system of difference equations for ψ and ϕ has to have zero determinant because there are nontrivial solutions of the homogeneous problem. Therefore, it is necessary that the system of difference equations be compatible. Note that for the ϕ equation, compatibility also involves the Neumann boundary condition. When the discrete variational principle is used, the difference equations are derived in a conservation form which not only guarantees compatibility but also suggests that they can be solved when only weak solutions of the continuous version of the problem are presumed to exist. The variational principle also provides a consistent derivation of the difference equations at the magnetic axis, which are complicated because of the singularity of the coordinate system there.

Once a discretization for the energy integral is chosen, all the corresponding difference equations and boundary conditions are uniquely defined. There is a strong analogy here to the finite element method, but instead of constructing an interpolation set for the dependent variables, we just write down a difference approximation to the integrand directly in terms of the nodal values of the unknown functions.

An exception must be made for the free surface equation. It may be viewed as a first-order partial differential equation for the function $g(u, v)$, although its coefficients also depend on the magnetic fields in the plasma and vacuum regions. Special attention is required to define a numerically stable iterative scheme for it. The continuous equation for g was derived after integration by parts under the assumption that the vacuum and plasma equations were already satisfied. This led to an equation on the free surface alone. If we use the variational principle, however, there will be contributions from the plasma and vacuum regions representing linear combinations of the operators in the interior, for in the discrete case they cannot be integrated out. The resulting difference equations are strongly coupled to the rest of the system, and it would be very difficult to design a stable iterative scheme for them. Therefore, we prefer to view the free boundary equation as a first-order partial differential equation for g and simply introduce a second-order accurate difference approximation to the continuous version $M(g) = 0$ given in Section 2.9.

An iterative scheme for the minimization of E is arrived at by formulating a suitable finite difference approximation to the partial differential equations for paths of steepest descent that we expressed in terms of the artificial time parameter t. For this purpose, the right-hand sides, which come from the Euler equations, are treated in the fashion that has just been described.

3.2 Difference Equations for the Plasma Region

As we have already observed, the choice of discretization for the energy integral determines the finite difference equations uniquely. However, there are many difference approximations of the integral that have the same order of accuracy and one must be aware of some general principles in order to make a good choice.

If the approximation to the integral is second-order accurate, the resulting difference equations will also be second-order accurate with the possible exception of those at the magnetic axis, where there is a singularity of the coordinate system. The accuracy at the magnetic axis will depend on the scaling of s, which can be like R or R^2. This is determined by the choice of the toroidal flux function $F_T(s)$. If $F_T(s) = s$, then $s \sim R^2$ and the equations at the magnetic axis will only be first-order accurate, while if $F_T(s) = s^2$, then $s \sim R$ and the equations will be second-order accurate. Note that we can adjust one of the fluxes arbitrarily by changing coordinates, whereupon the choice of the rotational transform $\mu(s)$ determines the right expression for the other.

For a fixed order of accuracy, there are two factors that must be taken into account. We want to minimize the truncation errors within the given order of accuracy, and the resulting iterative scheme must be numerically stable.

We start by defining on the cube

$$\Omega: 0 \leq s \leq 1; \quad 0 \leq u \leq 1; \quad 0 \leq v \leq 1$$

a lattice with uniform mesh sizes h_s, h_u, h_v. We replace any function $f(s, u, v)$ by its nodal values

$$f_{ijk} = f(ih_s, jh_u, kh_v).$$

We then write an approximation to the integral for the plasma energy E_1 by summing over all boxes in the lattice a second-order accurate finite difference approximation to the integrand evaluated at the center of each of the boxes. Thus

$$E_1 = h_s h_u h_v \sum_i \sum_j \sum_k \frac{Q}{2LK\Delta} + h_s \frac{1}{\gamma - 1} \sum_i \frac{m^\gamma}{(h_u h_v \sum_j \sum_k LK\Delta)^{\gamma - 1}},$$

where

$$\Delta = \frac{\partial(r, z)}{\partial(s, u)} = HRR_s, \qquad K = 1 + \varepsilon r,$$

and

$$Q = (r_u^2 + z_u^2)\psi_v^2 + (r_v^2 + z_v^2 + L^2 K^2)\psi_u^2 - 2(r_u r_v + z_u z_v)\psi_u \psi_v.$$

As a first general principle to guide us in the choice of the finite difference approximation, we want to decouple the different terms as much as possible. This will, in turn, lead to difference equations involving fewer numbers of mesh points. It will enhance the resolution of the scheme by minimizing truncation errors and will also result in a more stable numerical scheme.

For example, it would be a poor choice to approximate Q at the center of each box by writing down approximations there for each of the factors r_u^2, z_u^2, ψ_v^2, ... separately. This would result in coupling the values of ψ_{ijk} to those of r_{ijk} and z_{ijk} on three different flux surfaces corresponding to the indices $i - 1$, i, and $i + 1$. Instead, since Q involves only u and v derivatives, we approximate it at the center of a box by taking the average of values at the centers of two opposite faces, say at i and $i + 1$. More precisely, we put

$$\left(\frac{Q}{2KL\Delta}\right)_{i+\frac{1}{2}, j+\frac{1}{2}, k+\frac{1}{2}} = \frac{\left(\dfrac{Q}{K}\right)_{i, j+\frac{1}{2}, k+\frac{1}{2}} + \left(\dfrac{Q}{K}\right)_{i+1, j+\frac{1}{2}, k+\frac{1}{2}}}{4L\Delta_{i+\frac{1}{2}, j+\frac{1}{2}, k+\frac{1}{2}}},$$

so that the values of ψ, r, and z appearing in each term Q/K will only be coupled to those in the same flux surface, and the only coupling in the s direction will come from the Jacobian Δ. In general, it is best to average over the largest possible collection of terms rather than to average each individual factor.

A similar principle is that when approximating the square of the first derivatives we should always square first and then average. For example, the formula

$$(\psi_u^2)_{i, j+\frac{1}{2}, k+\frac{1}{2}} = \frac{1}{2h_u^2} [(\psi_{i, j+1, k} - \psi_{i, j, k})^2 + (\psi_{i, j+1, k+1} - \psi_{i, j, k+1})^2]$$

is better than computing ψ_u at i, $j + \frac{1}{2}$, $k + \frac{1}{2}$ and then squaring. A similar formula is used for ψ_v^2, whereas the only reasonable second-order accurate approximation to $\psi_u\psi_v$ is given by

$$(\psi_u\psi_v)_{i,j+\frac{1}{2},k+\frac{1}{2}} = \frac{1}{4h_uh_v}[(\psi_{i,j+1,k+1} - \psi_{i,j,k})^2 - (\psi_{i,j+1,k} - \psi_{i,j,k+1})^2].$$

The same procedure yields for Δ the expression

$$\Delta_{i+\frac{1}{2},j+\frac{1}{2},k+\frac{1}{2}} = \frac{H_{j+\frac{1}{2},k+\frac{1}{2}}}{8h_s}[(R_{i+1,j,k}^2 - R_{i,j,k}^2)$$
$$+ (R_{i+1,j+1,k}^2 - R_{i,j+1,k}^2) + (R_{i+1,j,k+1}^2 - R_{i,j,k+1}^2)$$
$$+ (R_{i+1,j+1,k+1}^2 - R_{i,j+1,k+1}^2)].$$

The motivation for these finite difference formulas can be understood by considering the simple case of Laplace's equation in two dimensions, which involves the Dirichlet integral

$$\|\phi\|^2 = \iint(\phi_x^2 + \phi_y^2)dx\,dy.$$

Using a rectangular mesh and following the principle of squaring before averaging, we are led to the usual five-point formula for Laplace's equation. If instead we approximate the first derivatives at the center of each rectangle and then square, we are led to a five-point formula involving a mesh point together with the four points diagonally removed. The resulting difference equations allow for a discontinuous solution because values at adjacent points are not related through them. Thus squaring before averaging not only results in the right coupling, but is also necessary for the stability of iterative schemes. We might also point out that if we used the finite element method with rectangular elements and bilinear functions to interpolate for ϕ, we would get another unusual set of difference equations. The answer is, of course, that one should use triangular elements, but for a nonlinear problem in three dimensions they become very cumbersome.

Truncation errors can be reduced by observing that because of our coordinate system the functions R and g are slowly varying. The functions $r_2(u, v)$ and $z_2(u, v)$, which describe the shape of the outer wall, are known explicitly—for example, as trigonometric polynomials—so we can compute their first derivatives exactly. Then the truncation errors for the first derivatives will be bounded by derivatives of R, g, r_0, and z_0, which are generally smaller than those of r and z. Thus, for example, we discretize the formula

$$r_u^2 + z_u^2 = [(r_1 - r_0)^2 + (z_1 - z_0)^2]R_u^2 + [(r_1)_u^2 + (z_1)_u^2]R^2$$
$$+ [(r_1 - r_0)(r_1)_u + (z_1 - z_0)(z_1)_u](R^2)_u$$

instead of writing a difference approximation in terms of r and z directly. The coefficients in square brackets are functions of u and v alone, and they are computed at the center of each rectangle by using the exact values of r_2, z_2, $(r_2)_u$, and $(z_2)_u$. Expressions for the terms R_u^2, R^2, and $(R^2)_u$ on each flux surface are obtained by squaring before averaging. Similar principles are followed to approximate the remaining contributions to Q and Δ.

Finite difference equations are found by setting all the partial derivatives of the discrete sum for E_1 with respect to the values of ψ, R, r_0, and z_0 at the mesh nodes equal to zero. As an example, the discrete equation for ψ is given in the terminology of Fig. 3.1 by

$$L_1(\psi)_0 = \frac{1}{2Lh_v^2} \left[(E_9 + E_{10})(\psi_2 - \psi_0) - (E_{11} + E_{12})(\psi_0 - \psi_4) \right]$$

$$+ \frac{1}{2Lh_u^2} \left[(G_9 + G_{12})(\psi_1 - \psi_0) - (G_{10} + G_{11})(\psi_0 - \psi_3) \right]$$

$$- \frac{1}{2Lh_u h_v} \left[F_9(\psi_5 - \psi_0) - F_{10}(\psi_6 - \psi_0) - F_{11}(\psi_0 - \psi_7) \right.$$

$$\left. + F_{12}(\psi_0 - \psi_8) \right]$$

$$= 0,$$

where

$$E = \frac{1}{\Delta}\left(\frac{r_u^2 + z_u^2}{K}\right), \qquad F = \frac{1}{\Delta}\left(\frac{r_u r_v + z_u z_v}{K}\right),$$

$$G = \frac{1}{\Delta}\left(\frac{r_v^2 + z_v^2 + L^2 K^2}{K}\right).$$

This equation is in conservation form and only involves values of ψ on one flux surface. The u and v derivatives of the mapping and the factor K are also

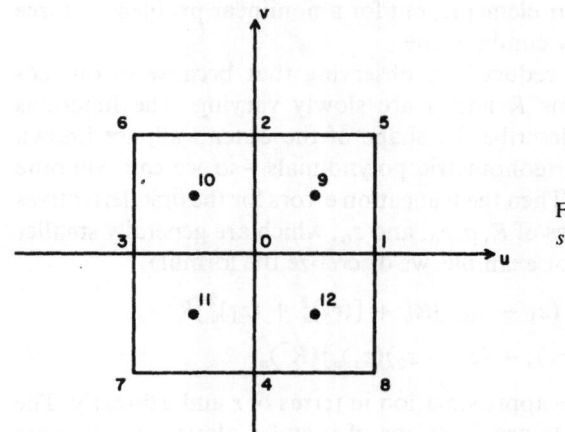

Fig. 3.1 Mesh on flux surface. $s = $ const.

computed on that flux surface. The equation is diagonally dominant because $F^2 \ll EG$. At the magnetic axis, where the derivatives r_u and z_u are equal to zero, it reduces to $(\partial/\partial u)(G\psi_u) = 0$ and can be solved explicitly using the periodicity conditions for ψ.

Details of all the difference equations for the plasma can be found from the listing of the computer code in the last chapter. Therefore, we will not go into them any further here.

3.3 Difference Equations for the Vacuum Region

For the vacuum, we proceed in a similar way by writing down a discrete approximation to the expression for Dirichlet's integral, given in Section 2.9. The domain Ω of the independent variables is the same as before, and we use the same type of uniform rectangular mesh with

$$\phi_{ijk} = \phi(ih_s, jh_u, kh_v).$$

To approximate the integrand we take, for example,

$$(a\phi_s^2)_{i+\frac{1}{2}, j+\frac{1}{2}, k+\frac{1}{2}} = \tfrac{1}{4}[(a\phi_s^2)_{i+\frac{1}{2}, j, k} + (a\phi_s^2)_{i+\frac{1}{2}, j+1, k}$$
$$+ (a\phi_s^2)_{i+\frac{1}{2}, j, k+1} + (a\phi_s^2)_{i+\frac{1}{2}, j+1, k+1}],$$

with similar formulas for $b\phi_u^2$ and $c\phi_v^2$, while for a typical mixed term we put

$$(d\phi_s\phi_u)_{i+\frac{1}{2}, j+\frac{1}{2}, k+\frac{1}{2}} = \tfrac{1}{2}[(d\phi_s\phi_u)_{i+\frac{1}{2}, j+\frac{1}{2}, k} + (d\phi_s\phi_u)_{i+\frac{1}{2}, j+\frac{1}{2}, k+1}].$$

The terms on the right-hand side are approximated in the usual fashion. The finite difference equation for ϕ as well as the boundary condition follow from setting to zero the derivatives of the discrete sum for the Dirichlet integral with respect to the values of ϕ_{ijk}, including those on Γ and C.

We must also compute the inductance matrix A in order to solve for the coefficients c_1 and c_2 determined by the flux constraints in the vacuum. This could be done by means of surface integrals, but it is more accurate to use the volume integral

$$a_{ij} = \iiint_{\Omega_2} \nabla\phi_i \cdot \nabla\phi_j \, dV,$$

which is stationary with respect to variations of ϕ. Thus an error in ϕ of order ε will lead to an error in a_{ij} of order ε^2. We approximate A by the same formulas that were used to derive the finite difference equation for ϕ.

3.4 Iterative Scheme for the Plasma Region

We define an iterative scheme to solve for ψ, R, r_0, and z_0 by writing down an explicit finite difference approximation to the partial differential equations for a path of steepest descent given in Section 2.10. Taking the equation for ψ as an example, we set

$$\frac{a_1}{(\Delta t)^2} (\psi_{i,j,k}^{n+1} - 2\psi_{i,j,k}^{n} + \psi_{i,j,k}^{n-1}) + \frac{e_1}{\Delta t} (\psi_{i,j,k}^{n+1} - \psi_{i,j,k}^{n}) = L_1(\psi)_{i,j,k},$$

with $t = n \Delta t$. The ratio $a_1/(\Delta t)^2$ must be chosen to satisfy the Courant–Friedrichs–Lewy stability condition [17]. For that, we can get a crude estimate by looking at the principal part of the differential operator and assuming that the coefficient of the mixed second derivative is zero. We freeze the coefficients of the equation and assume a solution of the form

$$\psi_{jk}^{n} = e^{i(\lambda n + \xi j + \eta k)}$$

to obtain the dispersion relation

$$\frac{a_1}{(\Delta t)^2} \sin^2 \frac{\lambda}{2} = \frac{1}{L} \left(\frac{E}{h_v^2} \sin^2 \frac{\eta}{2} + \frac{G}{h_u^2} \sin^2 \frac{\xi}{2} \right).$$

The stability requirement that λ be real for any real ξ and η leads to

$$\Delta t \le \left(\frac{La_1}{E/h_v^2 + G/h_u^2} \right)^{1/2},$$

which for $h_u^2/h_v^2 \le G/E$ can be estimated roughly by

$$\Delta t \le \left(\frac{a_1 \Delta_{\min}}{2L} \right)^{1/2} h_u.$$

We can now perceive one of the advantages of introducing second-order time derivatives in the method of steepest descent. Indeed, we have found that Δt scales like h_u, while without these derivatives similar analysis would show Δt to scale like h_u^2.

A special word must be said about these relationships near the magnetic axis. If we choose the scaling $s \sim R^2$, it turns out that the Jacobian Δ is essentially constant. However, if $s \sim R$, then Δ scales like s and $\Delta_{\min} \sim h_s$ at the flux surface nearest to the magnetic axis. This results in a change of the scaling for Δt. In the first case, the convergence will be faster but the truncation error near the axis will be correspondingly greater. In the second case, the convergence is slower but the truncation error is smaller. In our computer code, we scale R^2 like $s(1 + \alpha s)$, so that we can choose the parameter α to arrive at optimal results for each individual application.

In practice, one fixes the values of the coefficients a_i and finds the best time step Δt by making trial short runs on a crude mesh. Then one uses the scaling we have described to determine Δt for finer meshes.

Since r_0 and z_0 are functions of v alone, the Courant–Friedrichs–Lewy criterion is much less restrictive for the magnetic axis equations, given that the mesh has h_u and h_v of the same order of magnitude. This follows from the fact that the distances between mesh points in physical space are proportional to Rh_u and Lh_v in the u and v directions, respectively. Since for the most part $R \ll L$, the main restriction on Δt comes from the derivatives with respect to u, as we have already seen in the analysis of the equation for ψ. As a result, we can choose a_3 much smaller than a_1 and a_2.

The descent coefficients e_i are prescribed to be rather large numbers initially, say $e_i = 20$. After a few iterations they are adjusted automatically according to an acceleration formula of the form

$$e_i^n = \tau a_i \frac{1}{N} \sum_{j=n-N}^{n-1} \left| \frac{(F_t^j)}{F_i^j} \right|,$$

where

$$F(t) = \iiint \psi_t^2 \, dV.$$

Here N is the number of time steps over which an average is taken, and the parameter τ, which should exceed 1, is usually taken to be 2. In a typical run, the values of the e_i will decrease rapidly and then oscillate around some small number.

3.5 Iterative Scheme for the Vacuum Region

For the solution of Laplace's equation in the vacuum region, we use the standard successive over-relaxation scheme. This can also be interpreted as the discretization of an artificial time equation [18], with the modification that instead of adding a second-order time derivative one adds mixed space and time derivatives. The advantage is that with the mixed derivatives we do not need to store two different time levels, an economy which is important for three-dimensional problems.

The convergence properties of successive over-relaxation are similar to those of the second-order Richardson method. As in Section 2.11, we can improve the rate of convergence by allowing the relaxation factor ω to depend on artificial time according to the relation [18]

$$\omega(t) = \frac{2}{1 + e(t)h}.$$

Here $e(t)$ is computed in a manner similar to that for the plasma equations, and h is a typical mesh size.

Because of the minimax property of our variational principle for the potential energy E, we usually do several cycles of the vacuum iteration for each iterative cycle of the rest of the equations.

3.6 Iterative Scheme for the Free Boundary

As we said in Section 3.1, the free boundary equation can be viewed as a
first-order partial differential equation for the function g. In order to construct
a stable iterative scheme, we are guided by theory for the linearized equation.
We require a finite difference approximation that is second-order accurate in
space, since it is crucial to have good resolution. It is well known that the most
obvious explicit forward difference scheme is unstable. A two-step method like
"leap frog" is weakly unstable in the sense that there is an extraneous solution
whose growth rate does not depend on the mesh sizes h_u and h_v. For time-
dependent problems in which the solution is computed for only a short time
interval this is acceptable, because the initial conditions can be chosen so that
the coefficient of the unstable solution behaves like $h_u^2 + h_v^2$. However, to
solve the steady-state problem it is not satisfactory. The usual way of stabilizing
the scheme introduces an explicit dissipation term, but in order for the scheme
to remain second-order accurate this would involve an awkward fourth-
order difference operator.

We have overcome these difficulties by using a version of the Lax–Wendroff
method [29]. This is second-order accurate in both space and time. Moreover,
it has the property of not introducing truncation errors other than those
which come from the discretization of the space operator itself.

Let the partial differential equation for g be expressed in more detail as

$$e_4 g_t = M(g, g_u, g_v).$$

The idea of the Lax–Wendroff scheme is to compute the second time derivative g_{tt} from the equation and then substitute it into a Taylor series keeping
terms up to second order in time. This leads to a difference equation of the
form

$$g^{(n+1)} - g^{(n)} = \overline{\Delta t} M^{(n)} + \tfrac{1}{2}\overline{\Delta t}^2 (M_g M + M_{g_u} M_u + M_{g_v} M_v)^{(n)},$$

where $\overline{\Delta t} = \Delta t / e_4$ and $t = n\,\Delta t$. To be consistent with the interior equations
and to avoid excessive spreading of the difference equation, we evaluate the
finite difference approximation to M at the center of each of the rectangles
in the mesh. Let us introduce the nomenclature of Fig. 3.2. We define the values
of M, M_g, M_{g_u}, and M_{g_v} at each mesh node as an average of the values at the
four adjacent centers, so that, for example,

$$M_0 = \tfrac{1}{4}(M_1 + M_2 + M_3 + M_4).$$

Similarly, the derivatives M_u and M_v are specified as difference quotients of
M, such as

$$(M_u)_0 = \frac{1}{2h_u}(M_1 + M_2 - M_3 - M_4).$$

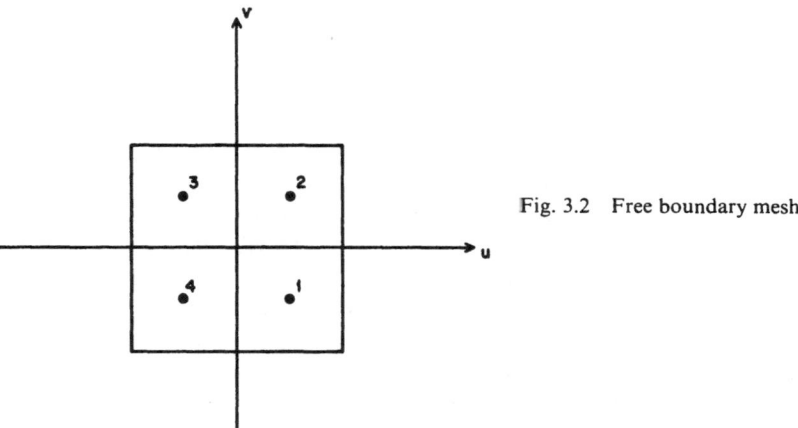

Fig. 3.2 Free boundary mesh.

We can view the right-hand side of the difference equation for g as an average of M with weights dependent on M_g, M_{g_u}, and M_{g_v}, which are evaluated by means of explicit differentiation. Because we have done the averaging on M itself, the steady-state solution satisfies $M = 0$ at the center of each rectangle. It is significant that no superfluous truncation error has been introduced.

The discrete approximation to M is done in the same manner as the approximation for the interior equations. All the derivatives involved are differenced within the free surface itself with the exception of R_s, which appears in the expressions for the plasma magnetic field and the pressure. For this term, we use a second-order accurate three-point formula.

From this discussion, it becomes clear why we have chosen to use Hadamard's formula for the free boundary equation rather than to derive an approximation from the discrete variational principle. It would be extremely difficult to arrive at a satisfactory iterative scheme if we had to consider all the interior terms instead of working with quantities on the free surface alone. It is important that the first-order equation for g is of the hyperbolic type, so that for numerical stability Δt scales like h_u and h_v, even though no second time derivative has been added. The coefficient e_4 is selected both to meet the Courant–Friedrichs–Lewy stability condition and to maintain descent. It is independent of the artificial time t.

3.7 Remarks About the Method

Because we are not only trying to solve the magnetostatic equations but also want to determine if the solution is stable, we have to distinguish between mathematical instabilities of the numerical method and physical instabilities of the equilibrium. This is accomplished by monitoring the behavior of the energy E as a function of the artificial time parameter t. Physical instabilities make E decrease indefinitely with increasing t, whereas numerical instabilities tend to make E either oscillate or grow.

The energy inequality $\delta E < 0$ can be lost numerically when the Courant–Friedrichs–Lewy criterion is violated or when the descent coefficients e_i are too small. In the first case, the iteration diverges rapidly and a smaller time step Δt must be used. In the second case, the growth rate λ of the artificial time equations has a large imaginary part that makes the energy and the residuals oscillate as functions of t. This can be avoided by using larger values for the constant τ, which appears in the formula for e_i given in Section 3.4.

In addition to the energy itself, there are several other physical and geometric quantities that we use to analyze the solution. The most important are the Fourier coefficients of R, r_0, z_0, and g, which determine the plasma displacements as functions of t, and the Jacobian D of the coordinate transformation as well as the residuals in the equations, which measure the magnetostatic forces.

When a solution is unstable, we find that one or more of the Fourier coefficients grows with increasing t, and the growth rate λ can be computed. As the instability develops, the unbalanced forces given by the residuals also grow, signifying that the plasma is drifting away from equilibrium. At the same time, we find that the Jacobian D becomes small until finally the procedure breaks down when D becomes negative, which is a mathematical indication of the plasma hitting the wall. Throughout the whole process δE is negative and increasing in absolute value. As we explained in Section 2.11, there will be a linear relation between the growth rate λ as computed by the code and the physical growth rate ω, with some additional empirical data needed to determine the constant of proportionality. In this sense, we can think of the method as providing information about the evolution of the system in physical time.

If the solution is stable, all the Fourier coefficients and geometrical quantities are convergent, with δE converging to zero through negative values and all the residuals converging to zero, which indicates force balance. When a vacuum region is present, the energy inequality may be lost toward the end of the process due to truncation errors arising from the free boundary equation and due to the minimax property of the variational problem. The equations inside the plasma were derived by minimizing the discrete expression for E_1 as a function of the nodal variables, and consequently truncation errors there will not affect the energy inequality, but the free surface equation was obtained simply by discretization of Hadamard's variational formula; therefore, it will be affected by trunction errors. There is no danger of confusion, however, since in the case we have cited all the quantities including the residuals are converging, while if the energy inequality is lost through numerical instability the iteration scheme diverges rapidly.

The effect of finite mesh size is to introduce truncation errors that we can liken to an anisotropic artificial viscosity. For the plasma field this is comparable to the effect of finite Larmor radius and tends to make equilibria appear more stable in the calculations than they are for the continuous magnetohydro-dynamic model. The discrete variational principle, like the finite element method, restricts the class of admissible functions in the minimization of E, which results in constraints that raise eigenvalues and stabilize equilibria. Because of the minimax character of the variational principle, the errors in the vacuum field have an opposite, destabilizing effect. Theoretically, the truncation error can always be assessed by refining the mesh, but limitations on computer capacity for large-scale, time-dependent calculations in three dimensions restrict what can be achieved in practice.

The question of existence of solutions to the fully three-dimensional equilibrium equations remains open. However, the existence of solutions to the discrete problem is a strong indication of the existence of weak solutions to our formulation of the continuous problem, for the success of the computations is in large part due to the presence of artificial viscosity terms coming from the truncation errors. There is no contradiction here with nonexistence arguments, which presuppose regularity properties of the solution [20].

A final word must be said to the effect that while theory provides us with a good motivation on which to base our numerical method, the ultimate test of the validity and correctness of what has been done lies in the performance of the computer code. The difficulties stemming from the nonlinearity and nonstandard type of the magnetostatic equations in three-dimensional space are too great to allow for a rigorous mathematical proof.

3.8 Iterative Schemes for Elliptic Equations

A simple and powerful tool for analyzing iterative schemes for elliptic partial differential equations is their interpretation as discrete approximations of time-dependent equations. Thus by writing down appropriate artificially time-dependent equations we can arrive at a variety of iterative schemes of practical interest.

To exhibit the principles involved, let us consider the elementary example

$$L(\phi_j) = \frac{\phi_{j+1} - 2\phi_j + \phi_{j-1}}{2} = 0, \qquad \phi_1 = a, \qquad \phi_n = b,$$

of a discrete approximation of Laplace's equation $\Delta\phi = 0$ in one dimension with the boundary conditions $\phi(0) = a$, $\phi(1) = b$. The simplest iterative scheme is given by

$$\phi_j^{n+1} = \tfrac{1}{2}[\phi_{j+1}^n + \phi_{j-1}^n] = \phi_j^n + L(\phi_j^n).$$

If we divide by $h^2/2$ and set $\Delta t = h^2/2$, this relation can be interpreted as a finite difference approximation to the heat equation

$$\phi_t = \phi_{xx}$$

on the lattice $x = jh, t = n\,\Delta t$. Such a procedure corresponds to the method of steepest descent for minimization of Dirichlet's integral $\int \phi_x^2\,dx$. Convergence to the steady-state solution occurs at the exponential rate $e^{-\lambda t}$, where λ stands for the lowest eigenvalue of

$$\Delta\phi + \lambda\phi = 0$$

with the requirement $\phi = 0$ at $x = 0, 1$. According to this estimate, the number of iterations needed to achieve a prescribed degree of accuracy grows like the reciprocal of h^2.

A slightly better scheme, known as the Gauss–Seidel method, is given by

$$\phi_j^{n+1} = \tfrac{1}{2}[\phi_{j+1}^n + \phi_{j-1}^{n+1}] = \phi_j^n + \tfrac{1}{2}[\phi_{j+1}^n - 2\phi_j^n + \phi_{j-1}^{n+1}].$$

By setting $\Delta t = h^2$, this can be interpreted as a finite difference approximation to $h\phi_{xt} + \phi_t = \phi_{xx}$. It converges about twice as fast as the previous scheme. The successive over-relaxation method is obtained by introducing a relaxation factor $\omega > 1$ to increase the weight of the correction given by the second term on the right. More precisely, it is defined by the formula

$$\phi_j^{n+1} = \phi_j^n + \frac{\omega}{2}[\phi_{j+1}^n - 2\phi_j^n + \phi_{j-1}^{n+1}],$$

which in turn can be interpreted as a discrete approximation to the hyperbolic equation [18]

$$\phi_{xt} + e\phi_t = \phi_{xx}$$

by setting $\omega = 2/(1 + eh)$ and $\Delta t = h$. Here the best choice for e and the convergence rate are also related to the lowest eigenvalue of $\Delta\phi + \lambda\phi = 0$. A major improvement comes from the fact that $\Delta t = h$, so now the number of iterations grows like the reciprocal of h instead of h^2. For elliptic equations in general, however, the theoretically optimal value for ω cannot usually be employed because it makes the iterations diverge in practice. Thus one is confined to a slower convergence rate. However, allowing the relaxation factor ω to depend on t by choosing e in the manner we have indicated in Section 3.4 enables us to attain the best choice for ω and correspondingly improve the convergence rate.

We can carry the procedure a step further with the aid of the fast Fourier transform. Consider the equation

$$L(\phi) = \frac{\partial}{\partial x}(a(x, y)\phi_x + b(x, y)\phi_y) + \frac{\partial}{\partial y}(b(x, y)\phi_x + c(x, y)\phi_y) = 0$$

in the square

$$\Omega: 0 \leq x \leq 1; \quad 0 \leq y \leq 1,$$

with ϕ prescribed on the boundary. This is the Euler equation corresponding to minimization of the functional

$$E(\phi) = \tfrac{1}{2} \iint (a\phi_x^2 + 2b\phi_x\phi_y + c\phi_y^2)dx\, dy.$$

The first variation is given by

$$\delta E = - \iint L(\phi)\delta\phi\, dx\, dy.$$

Let the iterative scheme be defined, following Jameson [26], by

$$\Delta\phi_t = -L(\phi).$$

He finds that

$$E_t = \iint \phi_t \Delta\phi_t\, dx\, dy = - \iint (\nabla\phi_t)^2\, dx\, dy < 0,$$

which proves convergence. We analyze this scheme by interpreting it as an artificial time equation and discover that the usual stability analysis leads to $\Delta t \sim 1$. Therefore, we have gained an order of magnitude in the time step and the number of iterations needed to achieve a specified degree of accuracy has become independent of the mesh size. The only difficulty is that at each time cycle we must invert the Laplacian, but this can be done rapidly using the fast Fourier transform.

It is clear that we can replace the Laplacian in Jameson's scheme by other elliptic operators with constant coefficients. The convergence rate will depend on how close an approximation to L the chosen operator is and on how fast that operator can be inverted. Such a procedure only becomes advantageous as the number of mesh points grows larger, with a break-even point at about 64 mesh intervals in each direction [3]. The method is readily applicable to the determination of the scalar potential ϕ of our magnetohydrodynamic problem because of the toroidal geometry. Since the vacuum field ought to be calculated accurately at each time cycle, the scheme would become useful if the plasma code were ever to be run employing a larger grid on some later-generation computer.

CHAPTER 4

Description of the Computer Code

4.1 Introduction

Here we describe how to use the computer code that is based on the theory developed in the previous chapters. The procedure for running the code is quite simple and only requires following the instructions presented in this chapter. However, an understanding of the theoretical framework is needed to determine the scope and limitations of the method, to choose the most efficient parameters, and to interpret the results correctly.

There are two kinds of input data. The first kind involves the physical and geometrical quantities of the problem, such as the equation of the outer conducting wall, the rotational transform and mass density distributions in the plasma region, and the poloidal and toroidal fluxes in the vacuum region. As an option, the vacuum region can be eliminated and the plasma region can be extended all the way to the outer wall. For that purpose, one can delete subroutines ASOR, ASBO, TSOR, CVl, CV, and TBO.

The second kind of data consists of parameters appearing in the numerical method, such as mesh sizes in space and time, coefficients of the equations for paths of steepest descent, and acceleration parameters for the iteration scheme. While the physical data is determined by the problem itself, a good choice of the numerical parameters is required for efficient use of the method. This involves theoretical estimates and trial runs of the program in more complicated cases.

Given the input data, the program minimizes the potential energy, and if the iteration is convergent it computes a stable equilibrium. The iteration can diverge either because there is no equilibrium or because the equilibrium is unstable. If the iteration does diverge, we must distinguish between two explanations, namely numerical instabilities and physical instabilities. The former are caused by a wrong choice of the numerical parameters and are characterized by a loss of the energy inequality $\delta E < 0$. A detailed analysis of how to avoid such difficulties is given later in the chapter. Physical instabilities are characterized by indefinite descent of the energy with the residuals of the equilibrium equations increasing. By means of a Fourier analysis of the solution, we can determine the particular mode that dominates such an instability.

When the iteration is convergent, we must keep in mind the effect of truncation errors in the minimization process. They tend to make the computed equilibria appear more stable than they really are. Often a small unstable eigenvalue is shifted so that it becomes stable. In order to assess the effect of the discretization errors, the mesh must be refined to see if the results remain unchanged.

In practice, there is a limitation on the mesh size we can use that is imposed by the availability of core space and computing time. This is a strong restriction on our method and is inherent in any three-dimensional calculation. However, there is an advantage in the simplicity of our formulation, which requires only the solution of two scalar equations. It allows us to introduce a finer mesh than would be possible if we used the fully time-dependent magnetohydrodynamic system.

The code, written in Fortran IV, has been run on the CDC 6600 at the Courant Mathematics and Computing Laboratory to study high β stellarator and Tokamak configurations. The results are presented in the next chapter. A typical run with NI = 16, NJ = 28, and NK = 28 mesh points in the s, u, and v directions, respectively, requires 300K octal core storage and about two hours of CP time. Using the CDC 7600 at Lawrence Berkeley Laboratory we have been able to run cases with finer meshes. We have used the FTN compiler with OPT = 2, obtaining a significant increase in speed over the RUN compiler.

Results are printed out and plotted on a Calcomp plotter using a slightly modified Calcomp plotter package. Users must adapt the plotting routines TPLOT and PLOTB to their own computing system. The plots are essential to a thorough understanding of the results.

4.2 Input Data

The input data for the code consists of a deck of data cards, described below. The names and definitions of all input parameters are listed in a glossary in Section 4.4. Odd cards are dummies that are used to identify input variables on the even cards (see page 72).

Card 2 ε, r_a, Q_L, NRUN—FORMAT(3F10.5,I5): The unit of length is equal to the average radius of a cross section of the outer wall. In these units, ε is the inverse of the major radius of the torus. The parameter r_a is the ratio of the free boundary radius to the wall radius and it is set equal to one when no vacuum region is present.

Q_L denotes the number of periodic sections into which the torus is divided. If $\varepsilon > 0$, the numerical solution of the problem is restricted to one such periodic section. For example, $\varepsilon = .02$ and $Q_L = 24$ means that the ratio of the major radius of the torus to the wall radius is 50 (since the wall radius is 1 by definition) and there are 24 periodic sections in the full torus. The computation is restricted to one section. In general, the toroidal angle is given by $\theta = 2\pi v/Q_L$, and since $0 \leq v \leq 1$, the period in θ is $2\pi/Q_L$. If a computation over the full torus is desired, it is sufficient to set $Q_L = 1$. If $\varepsilon = 0$, the major radius is infinite. This corresponds to the case of a straight cylinder, and Q_L becomes the length of the cylinder.

NRUN is a run identification number which appears on the printed and plotted output.

Card 4 $\Delta_0, \Delta_1, \Delta_2, \Delta_3, \Delta_{10}, \Delta_{20}, \Delta_{30}$—**FORMAT(7F10.5)** and **Card 6** Δ_{22}, Δ_{33}—**FORMAT(2F10.5)**: These parameters determine the shape of the outer wall according to the formulas

$$r_2(u, v) = r_b \cos U + \Delta_1 \cos V - \Delta_2 \cos (U - V),$$

$$z_2(u, v) = r_b \sin U + \Delta_1 \sin V + \Delta_2 \sin (U - V),$$

where $U = 2\pi u$, $V = 2\pi v$ and

$$r_b = 1 - \Delta_0 \cos (V) - \Delta_3 \cos (3U - V) + \Delta_{10} \cos (U) + \Delta_{20} \cos (2U)$$
$$+ \Delta_{30} \cos (3U) + \Delta_{22} \cos (2[U - V]) + \Delta_{33} \cos (3[U - V]).$$

Note that if all the Δ's are zero, we have circular cross sections. $\Delta_0, \Delta_1, \Delta_2$, and Δ_3 have the usual meaning that occurs in the high β stellarator literature. Δ_{10}, Δ_{20}, and Δ_{30} describe axially symmetric deformations. Δ_{22} and Δ_{33} are perturbations whose toroidal periods are a half and a third of those associated with Δ_2 and Δ_3, respectively. They define helically symmetric deformations in a straight configuration.

We have chosen this description for the equation of the outer wall because it includes all the cases, for which results will be presented in the next chapter. Any other expression satisfying the periodicity conditions can be used instead by making appropriate changes in subroutines ASIN and SURF. Note that because the code has been written so as to minimize truncation errors, not only must the equations for $r_2(u, v)$ and $z_2(u, v)$ be given, but also their u and v derivatives.

Card 8 $n, p_0, \mu_0, \mu_1, \mu_2$—**FORMAT(5F10.5)**: These parameters describe the initial pressure and rotational transform distributions according to the formulas

$$p(R) = p_0(1 - R^2)^n,$$

$$\mu(R) = \mu_0 + \mu_1 R + \mu_2 R^2,$$

where $R = 0$ at the magnetic axis and $R = 1$ at the plasma boundary. From the pressure distribution we initialize the mass density function $m(s)$, which then remains fixed throughout the iteration. While the pressure distribution itself does not remain fixed, the final pressure should be reasonably close to its initial values. It is simpler to think of the data in terms of the pressure rather than the mass, which is not so relevant physically.

The rotational transform $\mu(s)$ is the ratio of the s derivatives of the poloidal and toroidal fluxes $F_P(s)$ and $F_T(s)$. We are free to choose one of these functions arbitrarily in the initialization of $\mu(s)$, which also remains fixed throughout the iteration. We want the initial values to be close to an equilibrium solution. Therefore we model the procedure on the solution for a straight circular cylinder. In that case, the solution depends on the radial coordinate alone, and the toroidal and poloidal components of the magnetic field are given by

$$B_T = \frac{F'_T(s)}{\Delta}, \qquad B_P = \frac{2\pi F'_P(s)Rr_a}{L\Delta},$$

where $\Delta = \partial(r, z)/\partial(s, u) = 2\pi RR_s r_a^2$. If max $|\mu_i| \leq .1$, as in a high β stellarator case, the code sets $F'_T(s) = (1 - 2p(R))^{1/2}\Delta$ so that $B_T^2 + 2p = 1$, whereas if max $|\mu_i| > .1$, as in a Tokamak case, the code sets $F'_T(s) = \Delta$ so that $B_T = 1$.

Note that the toroidal field is normalized to be near unity at $R = 1$ and that a rough estimate of β on the magnetic axis is $2p_0$. It is a good idea to prescribe the initial pressure and rotational transform distributions to be solutions of the equilibrium equation

$$\frac{\partial}{\partial R}\left(p + \frac{1}{2}B^2\right) + \frac{B_P^2}{R} = 0$$

for the straight cylinder with circular cross section. More general expressions for $p(R)$ and $\mu(R)$ can be coded in subroutine ASIN.

Card 10 α, AMP—FORMAT(2F10.5): The parameter α defines the mesh distribution in the radial direction. The initial values of R are related to the flux by the formula

$$R^2 = \frac{s(1 + \alpha s)}{1 + \alpha}.$$

A large value of α corresponds to equally spaced mesh points in the radial direction, whereas $\alpha = 0$ corresponds to mesh points distributed like R^2. Large α will decrease the truncation error at the magnetic axis but will require a smaller time step. The choice of α depends on the problem to be investigated. We usually put $\alpha = 0$.

AMP determines the amplitude of the initial perturbation used to study the stability properties of the equilibrium under consideration. Theoretically, the unstable modes should ultimately appear simply because they are excited by the noise due to roundoff errors, but it would take too many iterations for their amplitudes to grow to a level where they are easily detectable. In practice, it is more efficient to input specific perturbations of the initial solution which one expects to lead to instabilities.

Card 12   FUR(I), $I = 1, \ldots, 7$—FORMAT(7F10.5) and **Card 14   FUZ(I), $I = 1, \ldots, 7$—FORMAT(7F10.5)**: FUR and FUZ are the Fourier coefficients of the r and z coordinates of the magnetic axis. The initial values for the magnetic axis are given by the axially symmetric solution plus a perturbation

$$r_0(v) = r_0 + \delta r_0, \qquad z_0(v) = z_0 + \delta z_0,$$

where

$$\begin{aligned}
\delta r_0 = \text{AMP}[&\text{FUR}(1) + \text{FUR}(2) \sin (V) + \text{FUR}(3) \cos (V) \\
&+ \text{FUR}(4) \sin (2V) + \text{FUR}(5) \cos (2V) + \text{FUR}(6) \sin (3V) \\
&+ \text{FUR}(7) \cos (3V)],
\end{aligned}$$

with a similar expression for δz_0.

Note that all perturbations of the magnetic axis correspond to the poloidal mode number $m = 1$. We have, for example,

$$\delta r_0 = \text{AMP} \cos (V), \qquad \delta z_0 = \text{AMP} \sin (V),$$

corresponding to a helical perturbation usually referred to as the $m = 1, k = 1$ mode. FUR and FUZ can also be used to initialize the magnetic axis in an appropriate position when the configuration is not axially symmetric. Thus for a high β stellarator we should take

$$\text{FUR}(3) \text{ AMP} = \text{FUZ}(2) \text{ AMP} = \Delta_1$$

so that the magnetic axis has the same helical excursion as the outer wall.

Card 16   FUX(I), $I = 1, \ldots, 7$—FORMAT(7F10.5): These are the Fourier coefficients of the initial perturbation of the flux surfaces, where the perturbation is given by

$$\begin{aligned}
\delta = \text{AMP}[&\text{FUX}(1) \cos (V) + \text{FUX}(2) \cos (2U) + \text{FUX}(3) \cos (3U) \\
&+ \text{FUX}(4) \cos (2U - V) + \text{FUX}(5) \cos (3U - V) \\
&+ \text{FUX}(6) \cos (2[U - V]) + \text{FUX}(7) \cos (3[U - V])].
\end{aligned}$$

If there is a vacuum region present, we initialize the free boundary function by means of the formula $g(u, v) = g(u)[1 + \delta]$, where $g(u)$ is obtained from an axially symmetric calculation. This results in a perturbation of every flux surface. If there is no vacuum region, we perturb the radial function

$$R(s, u, v) = R(s, u)[1 + (1 - s)\delta],$$

where $R(s, u)$ is the axially symmetric solution. A more general perturbation can be prescribed by making changes in subroutine SURF.

Card 18   NI, NJ, NK, ASYE, ERR—FORMAT(3I5, 2E10.3): NI, NJ, and NK are the numbers of mesh points in the $s, u,$ and v directions in the plasma region. $I = 1$ corresponds to the magnetic axis and $I = \text{NI}$ corresponds to the plasma boundary. The mesh sizes are $h_s = 1/(\text{NI} - 1)$, $h_u = 1/\text{NJ}$ and $h_v = 1/\text{NK}$. Because of the periodicity conditions, the meshes in the u and v directions contain $\text{NJ} + 2$ and $\text{NK} + 2$ points, respectively. $J = 2$ corresponds to $u = 0$ and $J = \text{NJ} + 2$ corresponds to $u = 1$, and both represent the same point in physical space. All calculations are performed for $J = 2, 3, \ldots, \text{NJ} + 1$ with the values of the functions at $J = 1$ and $J = \text{NJ} + 2$ defined by periodicity conditions. A similar procedure is followed in the v direction. Note that when prescribing NI, NJ, and NK, the corresponding dimensions in the code must be NI, $\text{NJ} + 2$, and $\text{NK} + 2$.

If no vacuum region is present, a mesh given by $\text{NI} = 16, \text{NJ} = 28, \text{NK} = 28$ requires 300K octal words of storage on the CDC 6600. When a vacuum region is allowed, a mesh given by $\text{NI} = 10, \text{NJ} = 28, \text{NK} = 28$, and with $\text{NIV} = 10$ more points in the radial direction for the vacuum region fits into 300K words of memory. When deciding on the type of mesh to use, one must consider the distribution of mesh points in physical space rather than in s, u, v space. In general, we want to avoid a mesh where distances between mesh points in physical space in the different coordinate directions vary a great deal, since this would lead to difficulties in the numerical analysis.

ASYE defines a criterion for the maximum error in the solution of the axially symmetric equations at which the computation becomes three-dimensional. The first step in the code is the solution of the axially symmetric problem, which is then used to initialize the three-dimensional calculation. If the equilibrium problem is axially symmetric (i.e. independent of v), we can calculate the solution with the axially symmetric routine and then use the full three-dimensional program to test stability properties. In this case, a typical value of ASYE is .0001. When the equilibrium problem is three-dimensional, and especially when an axially symmetric solution does not exist, we usually proceed to the three-dimensional calculation after a single iteration of the axially symmetric routine. In this case, ASYE is set to 100.

ERR defines a criterion for the maximum error in the solution of the three-dimensional equilibrium equations at which the calculation terminates.

Card 20 a_1, a_2, a_3, Δt—**FORMAT(4F10.5):** a_1, a_2, and a_3 are the coefficients of the second time derivatives in the equations for R, ψ, and the magnetic axis. Typical values are $a_1 = a_2 = 1.0$ and $a_3 = .025$. Δt is the time step. After the a_i are chosen, Δt must be adjusted to satisfy the Courant–Friedrichs–Lewy stability condition (cf. [17]). $\Delta t = .02$ is a typical value for a mesh with $\text{NI} = 17$, $\text{NJ} = \text{NK} = 28$. The simplest way to determine the time step is to make trial runs with a crude mesh and then to use appropriate scaling (see Section 3.4).

If Δt is too large, the iterations will diverge quickly and a negative Jacobian will terminate the computation. When Δt is marginal, oscillations of the energy may occur accompanied by large values of the descent coefficients e_i. In both cases, Δt should be decreased until the energy inequality is recovered.

The best way to find the right time step is to make a short run of about 100 iterations with small descent coefficients $e_i = 1$. In this case, the equations are not dissipative, and if the Courant–Friedrichs–Lewy criterion is satisfied, the residuals should oscillate without growing exponentially. We should not expect the iterations to converge nor the energy inequality to be satisfied because we have not provided adequate descent. The idea is that the ratio $a_i/(\Delta t)^2$ should be made large enough to insure numerical stability independently of the values of the descent coefficients. Otherwise the iterations might appear at first to be stable for large values of the e_i and then become unstable later when the descent coefficients are lowered by the acceleration scheme.

Card 22 e_1, e_2, e_3—**FORMAT(3F10.5):** e_1, e_2, and e_3 are the descent coefficients in the equations for R, ψ, and the magnetic axis. They are used as given during the first NAC iterations (see Card 26). After that they are determined by the acceleration scheme (see Section 3.4). Typical values are $e_1 = e_2 = 20$ and $e_3 = 2$. Note that they scale roughly like a_1, a_2, and a_3.

Card 24 τ_P, NE, NVAC—**FORMAT(F10.5,2I5):** τ_P is the proportionality constant between e_i and the time average of $|(F_i)_t/F_i|$ in the acceleration scheme for the equations in the plasma region. NE is the number of iterations over which the average is computed. Typical values are $\tau_P = 2$ and NE $= 50$.

The smoothness of the solution depends on the smoothness of the coefficients e_i as functions of time. Therefore, we want the e_i to vary slowly. For this, NE and τ_P must be large enough to avoid oscillations resulting from a large imaginary part in the dominant eigenvalue λ. On the other hand, if τ_P and NE are too large, the descent coefficients will not adjust fast enough to changes in λ.

NVAC positive indicates the presence of a vacuum region. If NVAC < 0, no vacuum region is present and the plasma region extends to the wall.

Card 26 NR, NZ, NT, NAC—**FORMAT(4I5):** NR and NZ indicate which Fourier coefficients of the r and z coordinates of the magnetic axis are fixed. NT gives the number of iterations for which those coefficients are held fixed. The values 1, 2, ..., 7 of the indices NR and NZ correspond to the terms 1, sin (V), cos (V), ..., cos $(3V)$, respectively. This procedure is useful for the study of unstable equilibria when one wants to initialize the plasma column with a given helical excursion or when one wants to examine the properties of a specific mode. For example, it allows the variables R and ψ to adjust while preserving the initial perturbation of the magnetic axis.

NAC denotes the number of the iteration after which the acceleration scheme for the descent coefficients takes effect. Acceleration should only start after an initial strongly nonlinear phase. Typical values are NR $=$ NZ $= 1$, NT $= 50$, and NAC $= 100$.

Card 28 IC, TLIM—**FORMAT(I5,F10.2):** IC gives the number of iterations after which, at most, 8 of 43 relevant quantities are printed. See also Card 32.

TLIM is a CP time limit after which the machine computation is terminated. One must allow for approximately 50 seconds of additional time to print and plot the final results.

Card 30 NIV, NV, NP, ω, τ_V, e_4, F_P^V, F_T^V—FORMAT(3I5,5F10.5): These parameters apply exclusively to the vacuum region, so the card should not be present if NVAC $<$ 0. NIV is the number of mesh points in the radial s direction. The number of mesh points in the poloidal and toroidal directions is the same as for the plasma region. NV is the number of iterations in the vacuum region for each free boundary iteration. NP is the number of iterations in the plasma region for each free boundary iteration. Typical values are NV $=$ 3 and NP $=$ 1, with the larger number of vacuum iterations due to the minimax property of the vacuum energy.

ω is the initial value of the relaxation factor for the vacuum equations. τ_V is the proportionality constant for the acceleration scheme to determine $\omega(t)$. Typical values are $\omega = 1.8$ and $\tau_V = 2$. e_4 is the descent coefficient for the free boundary equation. It must be chosen so that the ratio $e_4/\Delta t$ satisfies the Courant–Friedrichs–Lewy criterion. A typical value is $e_4 = 1$.

F_P^V is the poloidal flux in the vacuum region and F_T^V is the toroidal flux in the vacuum region. It is best to choose them so that the initial values of the fluid plus magnetic pressure $p + \frac{1}{2}B^2$ are continuous across the free boundary. Otherwise the solution will become very different from the initial input, and if the discontinuity is too large the free boundary iteration might diverge in the first few cycles. Using the straight circular cylinder as a first approximation and assuming that $p = 0$ at the free surface, we have $B_T = 1$ in the vacuum and, therefore, $F_T^V = \pi(1 - r_a^2)$. The corresponding poloidal field is given by $B_P = C/\sqrt{r^2 + z^2}$, which leads to

$$F_P^V = LC \int_{r_a}^1 \frac{1}{\xi}\, d\xi = LC \log \frac{1}{r_a}$$

with C to be determined from $B_P = C/r_a$ at the interface and

$$L = \begin{cases} Q_L, & \varepsilon = 0, \\ \dfrac{2\pi}{Q_L\varepsilon}, & \varepsilon \neq 0. \end{cases}$$

Card 32 PRINT1, PRINT2, ..., PRINT7—FORMAT(7A10) and **Card 34 PRINT8—FORMAT(A10):** The user can choose which parameters to print after each IC iterations. This is done by prescribing the relevant names exactly as they are given in subroutine PRNT. They include the maximum error for each of the equations, the increment of energy, the descent coefficients, and the Fourier coefficients for the magnetic axis and flux surfaces. A maximum of eight different names can be prescribed, with each name left-adjusted. The names are self-explanatory, with M and K used for the poloidal and toroidal mode numbers. If NVAC > 0 they refer to the free boundary function $g(u, v)$, and if NVAC < 0 they refer to $R(s, u, v)$ for a fixed s near .5.

Card 36   NRA1, NRA2, NZA1, NZA2, MK1, MK2, MK3, MK4—FORMAT (8I5): The user can choose what to plot at the end of the run. See subroutine FPRINT for the seven possible choices of the parameters NRA and NZA specifying Fourier coefficients to be plotted. Two NRA's and two NZA's can be selected, as well as four combinations of the mode numbers MK. It is for these Fourier coefficients that growth rates are computed.

4.3 Printed Output

The first two pages of output are the input data itself, which identifies the run. On the next page the residuals, Jacobian, and increment of energy for the iterations of the axially symmetric equations are printed. After that the final results of the two-dimensional calculation are presented. They include the position of the magnetic axis, the plasma energy, and the pressure as a function of s. For NVAC > 0 the output also includes the origin of coordinates (r_3, z_3) for the vacuum region, the vacuum energy, the total energy, the toroidal and poloidal currents, the inductance matrix, and Fourier coefficients of the free boundary function $g(u)$. The results of the three-dimensional calculation in their dependence on the iteration number follow. The quantities printed are determined by choices made on Cards 32 and 34. After the last iteration, the final values of the energy and Fourier coefficients of the solution are given. The following page contains the average on each flux surface of physical quantities such as the magnetic field, current, rotational transform, pressure, and plasma β as functions of the average radius. Selections from a sample run are shown on pages 72–77.

We define the current by the formula

$$I(s) = \oint_{s=\text{const.}} B \cdot dx,$$

where the line integral is taken in the u and v directions for the toroidal and poloidal currents, respectively. Note that for the poloidal current $I(0) \neq 0$, therefore,

$$\int_{s \le s_0} J \cdot dS = I(s_0) - I(0)$$

on $v = $ const. The magnetic field has been computed over the full mesh and can be printed as a function of s, u, and v. Other quantities such as the current density can only be evaluated after some additional calculations.

For NVAC > 0 a further page of output gives the Fourier coefficients of the vacuum coordinate axis as functions of time. These are of interest because together with $g(u, v)$ they determine the location of the free boundary (see Section 2.9). A page with the Fourier coefficients of the magnetic axis follows. Then the Fourier coefficients of a flux surface appear. If NVAC > 0 they correspond to the free boundary function $g(u, v)$, and if NVAC < 0 they correspond to $R(s, u, v)$ for a fixed s. The column headings are the poloidal and toroidal mode numbers M and K, with each entry representing the square root of the sum of the squares of all amplitudes corresponding to the same values of M and K. Next, the descent coefficients are printed and, for NVAC > 0, the relaxation factors are also printed as functions of time. Finally, growth rates are printed.

4.4 Glossary

Below we have listed all input parameters, indicating card number, mathematical symbol, Fortran name, input format, range when applicable, and meaning. They appear in appropriate order, and within each card the first entry goes to the left with the others following in the format prescribed. Cards 32 and 34 specify the different quantities to be printed as functions of the iteration number. We have chosen a typical example; the complete list of possibilities is to be found in subroutine PRNT.

Glossary

Card No.	Mathematical Symbol	Fortran Name	Format	Range	Meaning
2	ε	EP	F10.5	$0 \leq \varepsilon < 1$	Inverse major radius.
2	r_a	RBOU	F10.5	$0 < r_a \leq 1$	Plasma-to-wall radius ratio.
2	Q_L	QLZ	F10.5		If $\varepsilon > 0$, number of periodic sections; if $\varepsilon = 0$, cylinder length.
2	NRUN	NRUN	I5		Run identification number.
4	$\Delta_0, \Delta_1, \Delta_2, \Delta_3,$ $\Delta_{10}, \Delta_{20}, \Delta_{30}$	DEL0,DEL1, DEL2,DEL3, DEL10, ...	7F10.5		Fourier coefficients of outer wall.
6	Δ_{22}, Δ_{33}	DEL22, DEL33	2F10.5		Fourier coefficients of outer wall.
8	n	XPR	F10.5	$0 \leq n$	Pressure distribution exponent.
8	p_0	P0	F10.5	$0 \leq p_0$	Maximum pressure.
8	μ_0, μ_1, μ_2	AMU0, ...	3F10.5		Rotational transform coefficients.
10	α	ALF	F10.5	$0 \leq \alpha$	Scaling constant.
10	AMP	AMP	F10.5		Perturbation amplitude.
12	FUR(I), $I = 1, \ldots, 7$	FUR(I)	7F10.5		Fourier coefficients of r coordinate of magnetic axis perturbation.
14	FUZ(I), $I = 1, \ldots, 7$	FUZ(I)	7F10.5		Fourier coefficients of z coordinate of magnetic axis perturbation.
16	FUX(I), $I = 1, \ldots, 7$	FUX(I)	7F10.5		Fourier coefficients of flux surface perturbation. If NVAC > 0, free boundary function $g(u, v)$ is perturbed. If NVAC < 0, radius function $R(s, u, v)$ is perturbed.

Line	Symbol	Name	Format	Constraint	Description
18	NI,NJ,NK	NI,NJ,NK	3I5		Numbers of mesh points in the s, u, and v directions in plasma region.
18	ASYE	ASYE	E10.3		Axially symmetric error criterion.
18	ERR	ERR	E10.3		Three-dimensional error criterion.
20	a_1,a_2,a_3	SA1,SA2,SA3	3F10.5	$0 \le a_i$	Coefficients of the second time derivatives of R, ψ, and the magnetic axis.
20	Δt	DT	F10.5	$0 < \Delta t$	Time step.
22	e_1,e_2,e_3	SE1,SE2,SE3	3F10.5	$0 \le e_i$	Descent coefficients for R, ψ, and magnetic axis.
24	τ_p	SAFI	F10.5	$0 < \tau_p$	Proportionality constant for acceleration scheme in plasma region.
24	NE	NE	I5	$NE \le 100$	Number of points for computing average.
24	NVAC	NVAC	I5		If positive, vacuum region is present; if negative, plasma region extends to the outer wall.
26	NR, NZ	NR, NZ	2I5	$1 \le NR,NZ \le 7$	Indices of the magnetic axis Fourier coefficients to be fixed.
26	NT	NT	I5	$1 \le NT \le NAC$	Number of iterations for whch NR and NZ are fixed.
26	NAC	NAC	I5		Iteration number after which acceleration scheme starts.
28	IC	IC	I5	$1 \le IC$	Printout indicator.
28	TLIM	TLIM	F10.2		CP time limit.
30	NIV	NIV	I5	$6 \le NIV$	Number of mesh points in the s direction in vacuum region.
30	NV	NV	I5	$2 \le NV$	Number of vacuum iterations per boundary iteration.
30	NP	NP	I5	$1 \le NP$	Number of plasma iterations per boundary iteration.
30	ω	OM	F10.5	$0 < \omega < 2$	Relaxation factor.
30	τ_V	SAFV	F10.5	$0 < \tau_V$	Proportionality constant for acceleration scheme in vacuum region.

Glossary (*continued*)

Card No.	Mathematical Symbol	Fortran Name	Format	Range	Meaning
30	e_4	SE4	F10.5	$0 < e_4$	Descent coefficient for free boundary equation.
30	F_P^V, F_T^V	FV1,FV2	2F10.5		Total poloidal and toroidal fluxes in vacuum region.
32	AXIS ERR	AXIS ERR	A10		Maximum error of magnetic axis equation.
32	RO ERR	RO ERR	A10		Maximum error of R equation.
32	PSI ERR	PSI ERR	A10		Maximum error of ψ equation.
32	BOU ERR	BOU ERR	A10		Maximum error of free boundary equation.
32	VAC ERR	VAC ERR	A10		Maximum error of vacuum equations.
32	DEL ENER	DEL ENER	A10		Increment of energy
32	RMA CONS	RMA CONS	A10		Average r coordinate of magnetic axis.
34	ZMA CONS	ZMA CONS	A10		Average z coordinate of magnetic axis.
36	NRA1	NRA1	I5	$1 \leq NRA1 \leq 7$	Plot parameter for Fourier coefficients.
36	NRA2	NRA2	I5	$1 \leq NRA2 \leq 7$	Plot parameter for Fourier coefficients.
36	NZA1	NZA1	I5	$1 \leq NZA1 \leq 7$	Plot parameter for Fourier coefficients.
36	NZA2	NZA2	I5	$1 \leq NZA2 \leq 7$	Plot parameter for Fourier coefficients.
36	MK1	MK1	I5	$00 \leq MK1 \leq 33$	Plot parameter for mode numbers.
36	MK2	MK2	I5	$00 \leq MK2 \leq 33$	Plot parameter for mode numbers.
36	MK3	MK3	I5	$00 \leq MK3 \leq 33$	Plot parameter for mode numbers.
36	MK4	MK4	I5	$00 \leq MK4 \leq 33$	Plot parameter for mode numbers.

CHAPTER 5

Applications

5.1 Historical Development of the Code

The method we present here has been developed in several stages. As a first model, we considered a sharp boundary plasma with skin current [5,6]. Our first code was applied to the problem of finding stellarator equilibria. It was also used to analyze high β Tokamaks of low aspect ratio. There were several limitations to this approach. The model was too simple because it did not allow for magnetic structure and flux surface constraints within the plasma. Thus realistic pressure profiles could not be considered and internal modes could not be studied. Also, a fixed coordinate axis was used that limited the plasma column to deviations from axial symmetry smaller than the plasma radius.

The second model we investigated was that of a diffuse plasma extending all the way to the outer wall [7]. This allows the pressure to go to zero at interior points, so that there may be a pressureless plasma region. However, no vacuum region or sharp boundary was included. The main application of the resulting code was to the equilibrium and stability of high β stellarators, including the Scyllac at Los Alamos and the Isar Tl-B at Garching.

Our final model is a combination of the previous ones [8]. It includes both magnetic structure in the plasma and a sharp free boundary. We have compared results from all three codes for similar cases and the agreement is good. This provides us with a consistency check, since the three codes were written independently.

5.2 Comparison with Exact Solutions

Straight cylinder with constant μ: As a first example, we consider a straight cylinder with circular cross sections and length L. The equilibrium quantities depend only on a radial coordinate $\overline{R} = \sqrt{r^2 + z^2}$, and a family of exact solutions can be found by integrating an ordinary differential equation. We have used several solutions to check our equilibrium theory, but here we only consider a simple case for which stability results are known explicitly (cf. [35]).

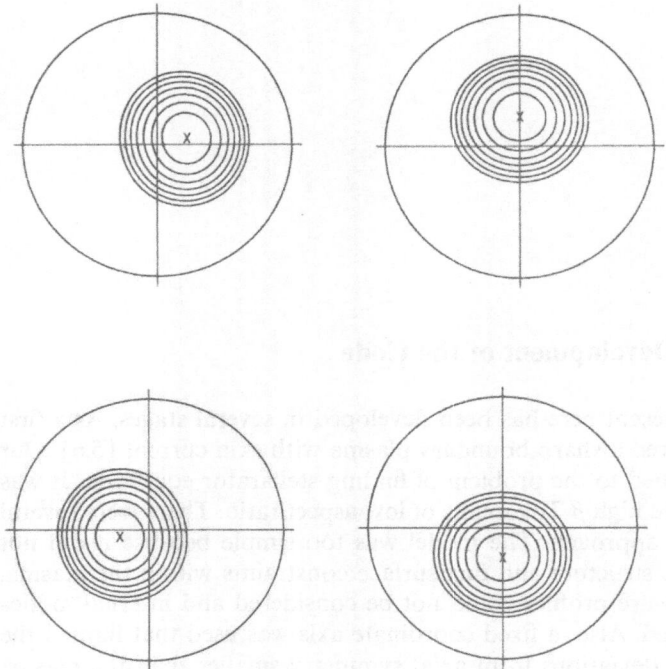

CROSS SECTIONS AT V= .04,.25,.54,.75, QLZ/2*PI= 1.50
MAJOR RADIUS INFINITE   MINOR RADIUS= 1.00

Fig. 5.1   Cross sections of flux surfaces.

Let the toroidal field $B_T = 1$ and let the rotational transform be a constant $\mu = \mu_0$ in the plasma region. The poloidal field B_P is given by $B_P = 2\pi\mu_0 \overline{R}/L$ there, which corresponds to constant current density. The condition for equilibrium gives

$$p(\overline{R}) = p_0 \left[1 - \left(\frac{\overline{R}}{r_a} \right)^2 \right],$$

where r_a is the plasma radius and $p_0 = (2\pi\mu_0 r_a/L)^2$. In the vacuum region we have $B_T = 1$ and $B_P = (2\pi\mu_0 r_a^2)/(L\overline{R})$.

In Figs. 5.1, 5.2, and 5.3 results for $L = 3\pi$, $\mu_0 = 1.6$, $r_a = .5$, and $r_b = 1$ are shown. We initialize the run with the equilibrium solution plus a perturbation of the free boundary. In Fig. 5.1, the flux surfaces in four cross sections of the plasma are shown. One can easily distinguish the $m = 1$, $k = 1$ mode. As artificial time evolves the potential energy develops an inflection point, and as the solution diverges away from equilibrium the residuals as well as the amplitudes of the unstable modes increase. This is typical of unstable equilibria. Notice that after the initial phase, the logarithms of the residuals increase

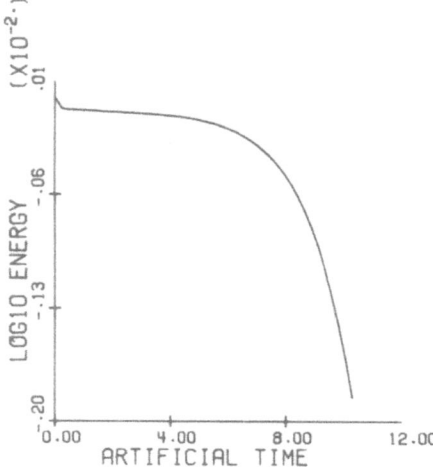

Fig. 5.2 Time behavior of energy and residuals.

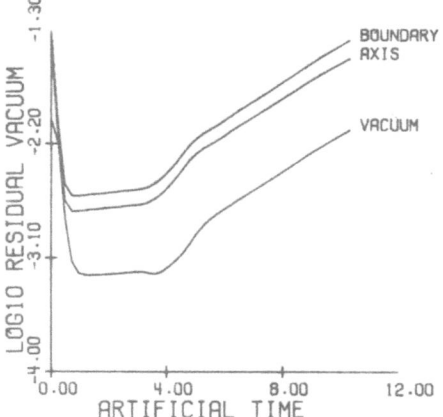

linearly with artificial time, indicating an exponential instability. The slope which is visible in the graphs of Fig. 5.2 agrees with the growth rate from the least-squares computation in the code for the $m = 1$, $k = 1$ mode. The ratios e_i/a_i of the descent coefficients to the acceleration coefficients are also shown. We can see that they decrease by a factor of 20 as artificial time increases. This results in a substantial acceleration of the iterative scheme. In contrast, for a stable equilibrium the energy converges to a minimum, while the residuals converge to zero and all the Fourier coefficients reach their equilibrium values.

If we continue the calculation of this case globally into a nonlinear regime, we find that after a while the instability stops growing and the residuals start converging, so that the plasma finds a new equilibrium position, shown in Fig. 5.4. The energy and the Fourier coefficients for the longer run are shown in

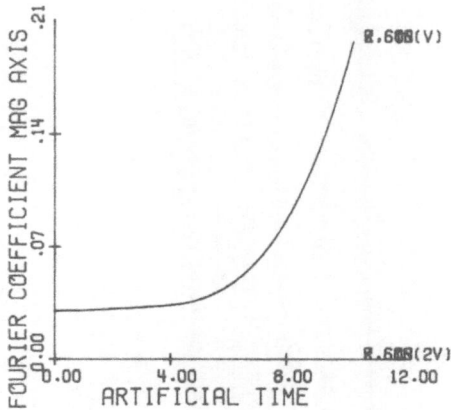

Fig. 5.3   Fourier coefficients and descent coefficients.

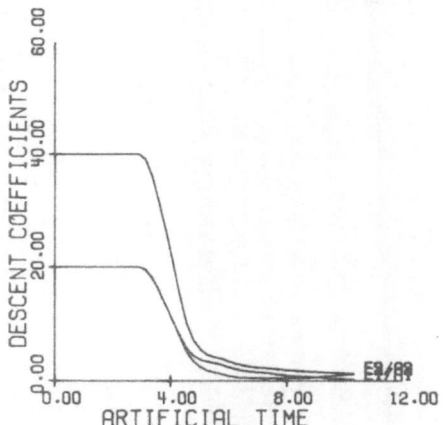

Fig. 5.5. It is seen that after the energy goes through an inflection point it converges to a minimum value, while the $m = 1, k = 1$ mode saturates. This is exactly what one would expect when a stable bifurcated solution exists [27]. Although this solution is not very remarkable physically because the plasma boundary is very near the outer wall, it does illustrate the nonlinear capability of the code.

In Section 2.11 we gave one method of computing unstable eigenvalues λ^2 for the artificial time equations, and we related them to the physical eigenvalues γ^2 of magnetohydrodynamics. A better way to estimate growth rates is to fit lines by least squares to logarithms of relevant Fourier coefficients. In Fig. 5.6, we show the dependence of such an approximation of the eigenvalues on the mesh size. Then, in Fig. 5.7, we compare them with the analytical results for the magnetohydrodynamic eigenvalues.

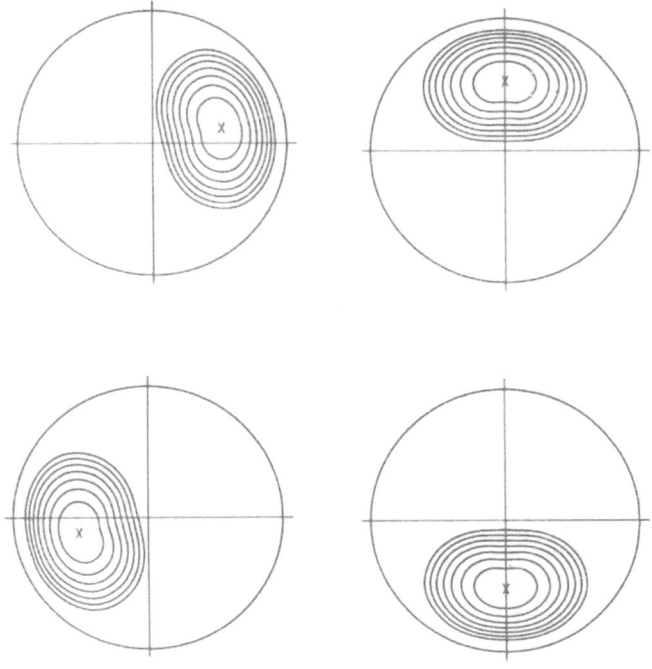

CROSS SECTIONS AT V= .04,.25,.54,.75, QLZ/2*PI= 1.50
MAJOR RADIUS INFINITE   MINOR RADIUS= 1.00

Fig. 5.4 Flux surfaces for bifurcated equilibrium.

To study eigenvalues, we fix the coefficients of the steepest descent equations as well as the other numerical parameters in the code. We then compute the growth rates λ for the $m = 1, k = 1$ free surface mode corresponding to different values of μ_0 and r_a. We can compare the analytical growth rate $\bar{\gamma}$ defined according to Shafranov [35] by

$$\bar{\gamma}^2 = \frac{4\pi\rho r_a^2}{B_P^2}\gamma^2 = 2\left[1 - \frac{1}{\mu_0}\right]\left[1 - \frac{1 - 1/\mu_0}{1 - r_a^2/r_b^2}\right]$$

with the computed growth rate $\bar{\lambda} = \lambda/B_P$, where the poloidal field B_P is evaluated at $\bar{R} = r_a$. Figure 5.7A shows the ratio $\bar{\gamma}/\bar{\lambda}$ as a function of r_a/r_b. We find that while $\bar{\gamma}$ alone varies by 15%, the ratio is nearly constant and varies only by 3%. Figure 5.7B represents the same ratio as a function of μ_0. Again it is nearly constant, and while $\bar{\gamma}$ varies by 60% the ratio varies only by 8%. Figure 5.7C shows $\bar{\gamma}$ and $c\bar{\lambda}$ as functions of μ_0, where c is the average value of the ratio $\bar{\gamma}/\bar{\lambda}$.

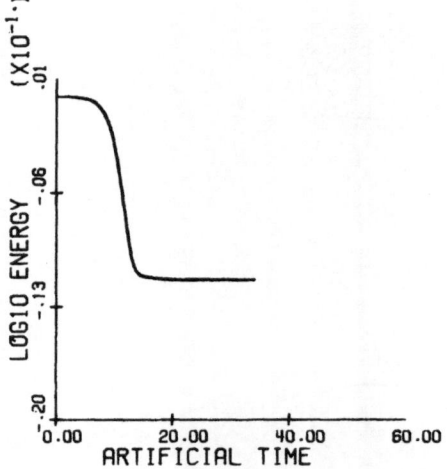

Fig. 5.5 Time plots for bifurcated equilibrium.

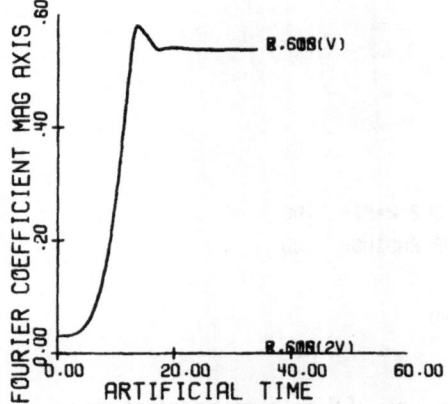

The conclusion to be drawn is that a dimensionless ratio of the growth rates with artificial and physical time can be found which is more or less independent of the rotational transform μ_0 and the compression ratio r_b/r_a. A linear stability theory of our artificially time-dependent equations could be worked out to find out more precisely how the growth rate λ depends on the numerical parameters of the problem. Only then could values for the growth rate γ in real time units be computed exactly.

Figure 5.8 shows the development of an $m = 2, k = 1$ mode. Note that the nonlinear behavior of the instability has the tendency to split the plasma in two. Globally, this would result in a splitting of the magnetic axis and, therefore, in a change in the topology of the flux surfaces, which is not allowed by our formulation. The growth rates of internal modes are much smaller in size than those for free boundary modes and, therefore, they are harder to detect. In

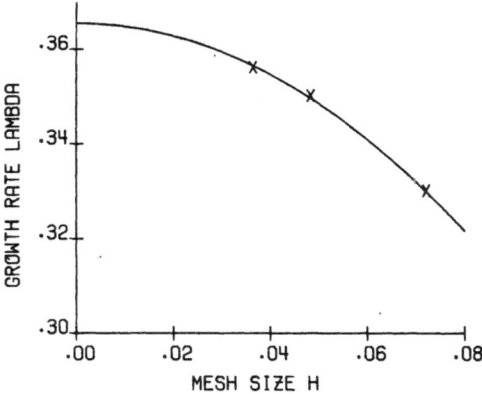

Fig. 5.6   Growth rate versus mesh size.

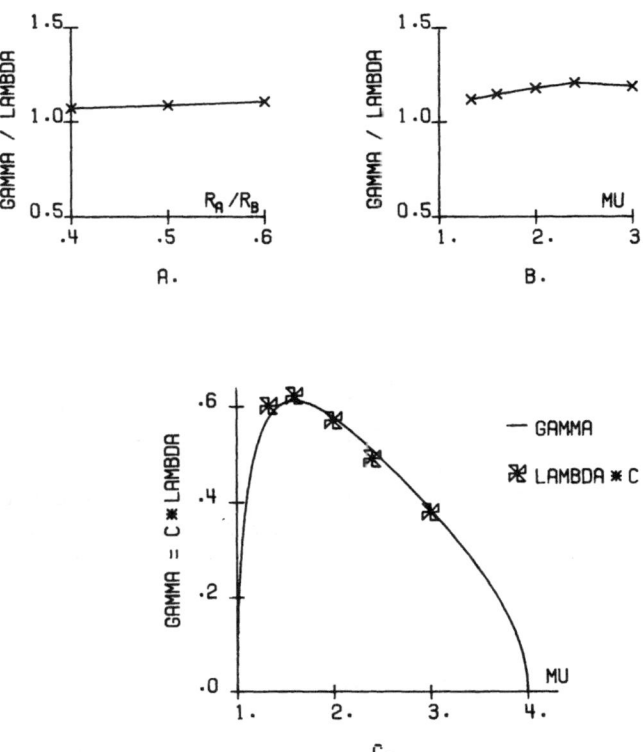

Fig. 5.7   Comparison of computed and exact growth rates.

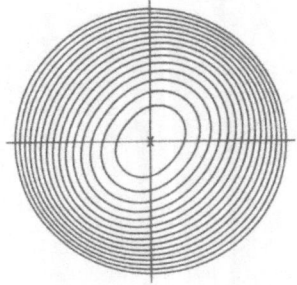

A. ARTIFICIAL TIME = 0 , V = .28

Fig. 5.8   Flux surfaces for $m = 2$ instability at two different times.

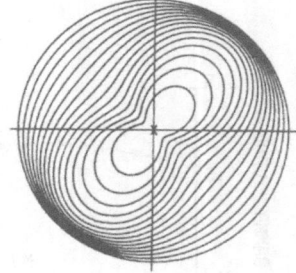

B. ARTIFICIAL TIME = 3 , V = .28

MU=.62, R_A=1, BETA=.9, 16X28X16 MESH POINTS

particular, it is easier to compute an $m = 1$, $k = 1$ instability than an $m = 2$, $k = 1$ instability. For the latter, the number of mesh points in the u direction is most important. Only high β cases are found to be unstable by the code for the kind of mesh sizes we are able to use.

The accuracy of the equilibrium computation can be checked by considering various exact solutions and seeing how well the code reproduces them. This has been done for several different versions of the screw pinch, and the results are satisfactory.

Axially symmetric solution: To validate the implementation of toroidal effects in the code, we have made a comparison with a family of exact solutions in the axially symmetric case. Now let r, θ, and z be cylindrical coordinates. The axially symmetric equilibrium equation for the flux function $\chi(r, z)$ is

$$\frac{\partial^2 \chi}{\partial z^2} + r \frac{\partial}{\partial r} \left(\frac{1}{r} \frac{\partial \chi}{\partial r} \right) = -r^2 \frac{dp}{d\chi} - I \frac{dI}{d\chi},$$

where $p = p(\chi)$ and $I = I(\chi)$ are given. If we choose

$$p(\chi) = p_0(1 - \chi),$$
$$I(\chi)^2 = l^2(1 - 4F_0\chi)$$

for $0 \le \chi \le 1$, the corresponding solution is

$$\chi(r, z) = F_0 l^2 z^2 + \frac{p_0}{8}(r^2 - l^2)^2.$$

We see that $\chi = 0$ corresponds to the magnetic axis $r = l, z = 0$, and we take $\chi = 1$ to be the equation of the outer wall. For the computation, we choose $l = 4$, $p_0 = \frac{1}{8}$ and $F_0 = \frac{1}{40}$. The outer wall is approximately an ellipse with major axis 1.58 and minor axis 1.04. Explicit integration shows that the rotational transform is

$$\mu(\chi) = \frac{2}{\sqrt{10}}\sqrt{\frac{4 - \chi}{1 - \chi/10}}$$

and the mass density function is

$$m(\chi) = 4\pi^2\sqrt{10}\left(\frac{1 - \chi}{8}\right)^{3/5},$$

where we have chosen $\gamma = \frac{5}{3}$.

We set

$$\chi = \frac{s(1 + \alpha s)}{1 + \alpha}$$

and compare the analytical solution with computations based on the code. Note that changes in the value of α alter the distribution of mesh points in the radial direction. We computed the solution for five meshes of between 9×14 and 33×56 points in the s and u directions. The results for the position of the magnetic axis with $\alpha = 0$ and $\alpha = 2$ are compared with the exact solution in Fig. 5.9A. We can see that for $\alpha = 2$ the rate of convergence is faster than $O(h)$, while for $\alpha = 0$ it is slower than $O(h)$. However, for the crude meshes the error for $\alpha = 0$ is smaller. In Fig. 5.9B, the maximum over $0 \le u \le 1$ of the difference between the analytic solution χ and the computed solution χ_h is shown for $s = \frac{1}{8}$ and $s = \frac{6}{8}$. Observe that the error is converging to zero faster for large values of s and for larger values of α. For $\alpha = 2$ and $s = \frac{6}{8}$ the rate of convergence is essentially $O(h^2)$, while for $\alpha = 0$ and $s = \frac{1}{8}$ it is $O(h)$. We also find that the energy E converges like $O(h^2)$. This confirms the fact that away from the magnetic axis the computational method is second-order accurate.

As we have already mentioned, an internal check of the code has been obtained by comparison of results from three different versions. Another type

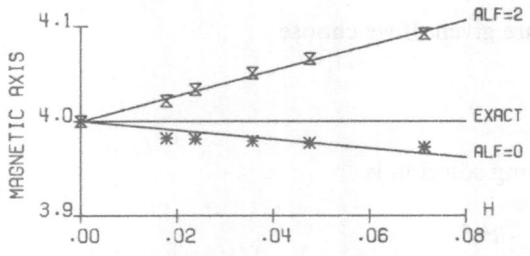

A.POSITION OF MAGNETIC AXIS VERSUS MESH SIZE

Fig. 5.9 Computed versus exact axially symmetric equilibria.

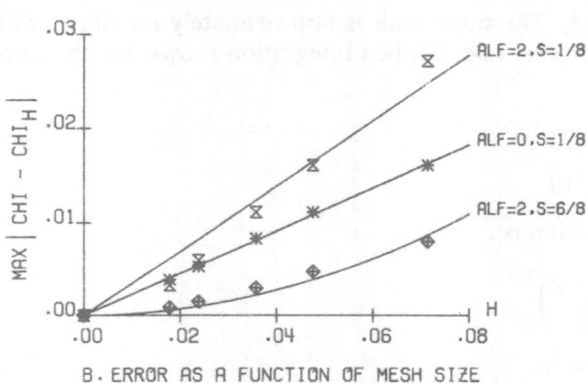

B. ERROR AS A FUNCTION OF MESH SIZE

of check can be made by comparing axially symmetric results from our two-dimensional routine with the corresponding fully three-dimensional calculations. This procedure has been a successful tool for validation of the code as well as for more elaborate stability analysis.

5.3 Unstable High β Stellarator Equilibria

An ideal field for application of the code we have described is the study of high β stellarators. On the one hand, our method gives better resolution for higher values of the plasma parameter β. On the other hand, the large computer resources required for calculations with three independent space variables and artificial time are only justified for genuinely three-dimensional problems such as are encountered with helical windings in toroidal geometry. Thus we have been motivated to perform extensive calculations modeling the Scyllac at the Los Alamos Scientific Laboratory and the Isar Tl-B at the Max Planck Institute for Plasma Physics in Garching [15,16]. The questions investigated were the existence of equilibria, especially in unstable cases, the estimation of the restoring force needed to offset outward toroidal drift, and a search for stabilized configurations.

For the study of high β stellarators, we work primarily with the reduced version

$$r_2 + iz_2 = r_b e^{iU} + \Delta_1 e^{iV} - \Delta_2 e^{-i(U-V)},$$
$$r_b = 1 - \Delta_0 \cos V + \Delta_{33} \cos 3(U - V)$$

of the outer wall formulas specified on Cards 4 and 6 of the input deck for the code. Our Δ's, which are associated with the perfectly conducting wall, are to be distinguished from the lower case δ's associated with the plasma column in the literature of the sharp boundary theory. They serve to define the various k, $l = 0, 1, 2, 3$ coils controlling the magnetic field in our model. The aspect ratio of the outer toroidal wall, which is the reciprocal of our parameter ε, was 50 for a typical version of the Scyllac and is 16 for the Isar Tl-B. The former had 45 periods and the latter 16.

For our model of a high β stellarator, we impose the flux constraints

$$\mu \equiv 0, \qquad F_P^V = 0,$$

which seem more appropriate than restrictions on the toroidal current. Most of the calculations are confined to one period of the device, which restricts the analysis of stability to modes with integral $k = 0, 1, \ldots$. In the examples to be described, we have used an average β of .7 and have taken the pressure profile initially to be either a Gaussian distribution or of the form

$$p(R) = p_0(1 - R^2)^8.$$

The compression ratio ranges from 2 to 5, and usually there is a large region of nearly pressureless plasma. Figure 5.10 shows the flux surfaces in four cross sections of one helical period for a run of the code modeling the Isar Tl-B. A mesh of $16 \times 24 \times 24$ points was used, with the free boundary half-way between the magnetic axis and the outer wall.

The existence of equilibria unstable to the $m = 1$, $k = 0$ mode can be established by making a series of runs in which the position of the magnetic axis is initialized at different distances $l + r$ from the main axis of the torus. If the plasma moves out with increasing artificial time when it is initialized far enough out, but moves in when it is initialized further in, then there is an unstable equilibrium that can be located by finding an intermediate value of r at which the plasma tends to move much more slowly. This is illustrated in Fig. 5.11. However, to obtain reliable results the average coordinates r and z of the magnetic axis must be held fixed for enough iterations to assure that other quantities, such as its helical excursion, have nearly reached their equilibrium values.

Figure 5.12 shows how erroneous conclusions can be drawn about equilibrium when the helical excursion is not initialized properly. In fact, the motion

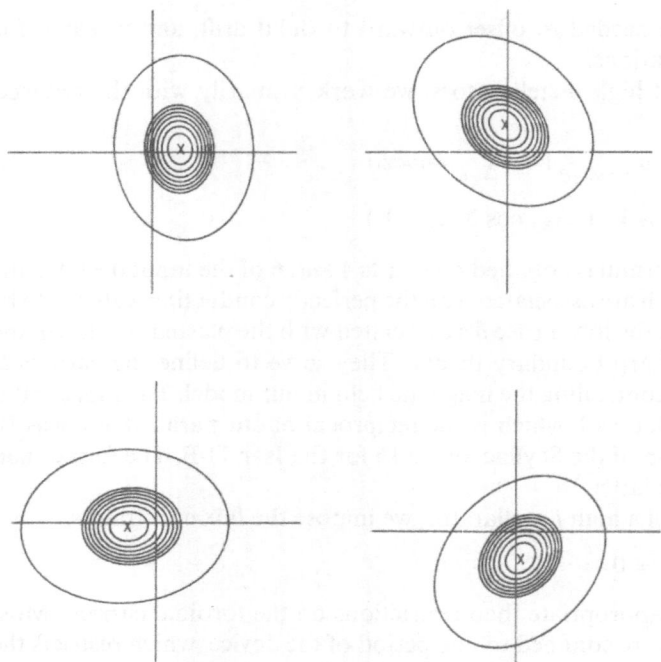

```
CROSS SECTIONS. AT V= .02,.27,.52,.77, 1/(EP*QLZ)= 1.00
MAJOR RADIUS= 15.94                    MINOR RADIUS= 1.00
```

Fig. 5.10   Cross sections for Garching Isar Tl-B.

of the plasma is seen to depend significantly on the initial excursion of the magnetic axis. This is borne out by experimental work on Scyllac.

The problem of unstable equilibrium and restoring force is illustrated for the Isar Tl-B in Fig. 5.13. Here the magnetic axis is initialized at the same location for every run, but the value of Δ_2 controlling the $l = 2$ field is varied. The equilibrium value of approximately $\Delta_2 = .14$, at which the plasma is seen to move most slowly, agrees reasonably well with the field that gave the best force balance in the experiment. A corresponding streak plot of the experimental data, obtained from reference [16], is presented in Fig. 5.14.

The calculations of unstable equilibrium have been verified by checking that the residuals measuring force balance are all relatively small. This is achieved by running the code with the average position of the magnetic axis held fixed at various locations and assessing the resulting errors in the axis equations.

In all cases where nonlinearity is significant, the estimates of restoring force calculated by means of the code turn out to be larger than corresponding

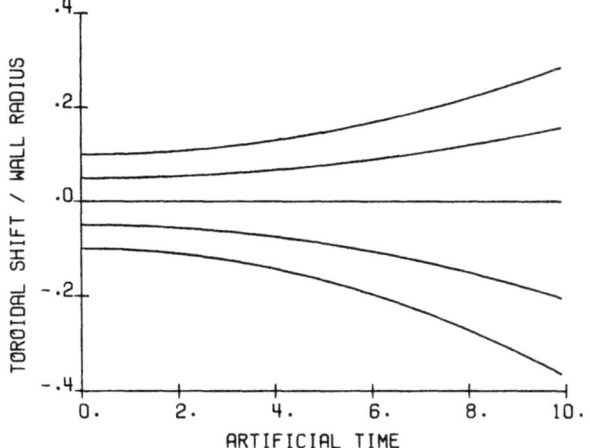

Fig. 5.11 Influence of initial displacement on Isar Tl-B.

Δ_0 =.0553, Δ_1 =.3647, Δ_2 =.1426, BETA =.7, AMP=.36

EP=.0627, NT=50, (8+8)X24X24 POINTS

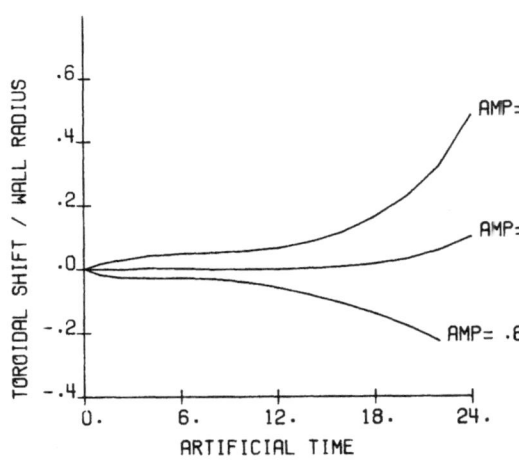

Fig. 5.12 Influence of initial helical excursion.

Δ_0 =.0553, Δ_1 =.3647, Δ_2 =.1426, BETA=.7

EP=.0627, NT=0, (8+8)X24X24 POINTS

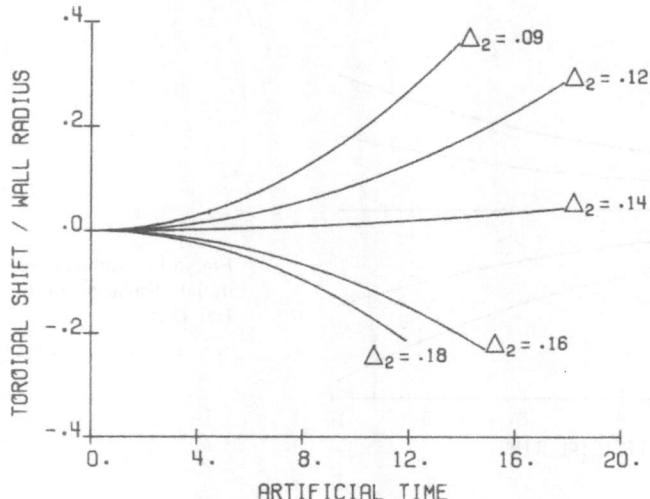

$\Delta_0 = .0553$, $\Delta_1 = .3647$, BETA=.7, AMP=.36

EP=.0627, NT=50, (8+8)X24X24 POINTS

Fig. 5.13 Streak plot of Isar Tl-B calculations.

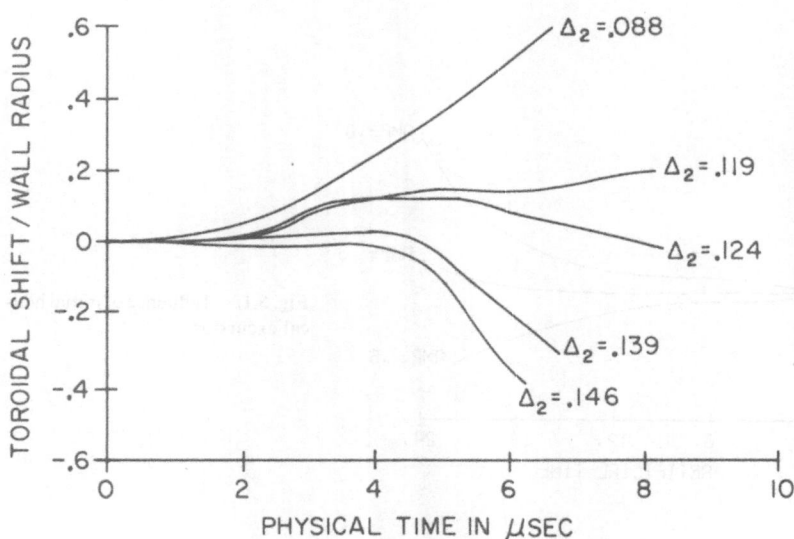

$\Delta_0 = .0553$, $\Delta_1 = .3647$, BETA = .7, EP = .0627

Fig. 5.14 Streak plot of Isar Tl-B experiment.

predictions from sharp boundary theory [34,37], which shows the force to be a linear combination of the products $\Delta_1\Delta_0$ and $\Delta_1\Delta_2$. Similar results were obtained numerically by Barnes and Brackbill [1,2]. These were confirmed experimentally at Los Alamos for an $l = 1, 2$ system with variable $l = 2$ coils. It was found that an increase of 20 % in the restoring force of sharp boundary theory could double the containment time of Scyllac, bringing it up to 50 μsec.

5.4 Triangular Cross Sections

The principal difficulty with high β stellarators is the $m = 1, k = 0$ instability, which appears as a translation of the whole plasma column toward the perfectly conducting outer wall. We have found that runs of the computer code are sensitive enough to this mode to give a realistic analysis of its behavior. It has been observed that truncation errors in the code tend to suppress instabilities associated with larger values of $m \geq 2$, except perhaps when there is a sharp boundary with a significant jump of the pressure p. This stabilizing effect, which is anisotropic and has special sensitivity to the poloidal mesh size h_u, is comparable to the physical effect of finite Larmor radius (see Section 3.7). One is led to the conclusion that for an appropriate mesh the code may provide a better simulation of physical reality than is expected from the classical magnetohydrodynamic theory.

It is our intention in this section to discuss the influence of the shape of the outer conducting wall on the $m = 1, k = 0$ instability. Runs of the code indicate that this mode is stabilized by a variety of noncircular cross sections having the same helical symmetry as the Δ_1 excursion. It suffices to present the results for the straight helically symmetric case $\varepsilon = 0$, with $\Delta_0 = \Delta_2 = 0$. The most effective stabilization seems to be associated with triangular cross sections, for which only Δ_1 and Δ_{33} differ from zero in our formulas for the wall.

Experimental work on this phenomenon is under way at the Max Planck Institute for Plasma Physics at Garching. Since the code has been used in the selection of parameters for the experiment, we shall describe calculations based on those parameters. Thus we take the compression ratio between 3 and 4 with $\beta = .7$, $L = 10\pi/3$ and $\Delta_1 = .5$. We allow r_a and Δ_{33} to vary in the intervals $.266 \leq r_a \leq .66$ and $0 \leq \Delta_{33} \leq .3$. The experimental configuration has a total of seven periods and a maximum duration of 20 μsec. For circular cross sections, the $m = 1, k = 0$ instability of the plasma is expected to show up within 7 μsec. Figure 5.15 displays the triangular wall, the vacuum region, and the plasma flux surfaces in four cross sections of one period of such an equilibrium that turned out to be stable for $r_a = .266$ and $\Delta_{33} = .3$. No pressureless plasma is present in this case.

In Fig. 5.16, two plots are shown of the trajectories of the average r coordinate of the magnetic axis corresponding to different initial values. The

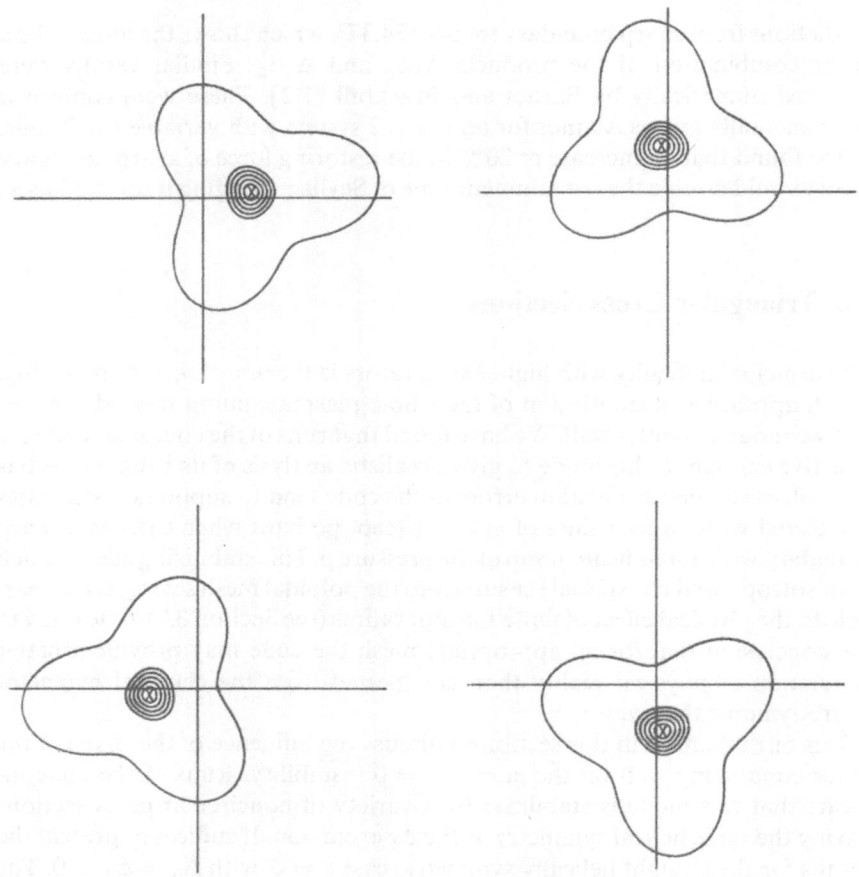

```
CROSS SECTIONS AT V= .02,.27,.52,.77, QLZ/2*PI= 1.66
MAJOR RADIUS INFINITE                    MINOR RADIUS= 1.00
```

Fig. 5.15 Triangular cross sections defining stable equilibrium.

first plot indicates how the trajectories diverge away from the equilibrium position $r = 0$ in the unstable case $\Delta_{33} = 0$ of circular cross sections. The second plot demonstrates that the equilibrium has become stable in the case $\Delta_{33} = .3$ of pronounced triangular cross sections. Here all the trajectories appear to converge to the same limit $r = 0$ as the artificial time increases. Growth rates of the $m = 1$, $k = 0$ mode can be estimated fairly well from the runs generating these diagrams. They vary slowly enough with mesh size to suggest that the stabilizing effect of the triangular cross sections is not just due

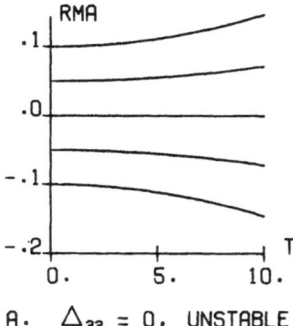

A. $\Delta_{33} = 0$, UNSTABLE

Fig. 5.16 Streak plots of unstable and stable equilibria.

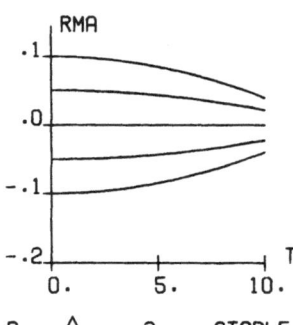

B. $\Delta_{33} = .3$,   STABLE

$\Delta_1 = .5$, BETA $= .7$, EP $= .0$, $R_A = .27$

to truncation error (see Fig. 5.17). The runs for the plots were performed on a mesh with NI + NIV = 20 and NJ = NK = 24 (see the description of Card 18 in Chapter 4).

Figure 5.18 displays the dependence of the stability on the magnitude of the coefficient Δ_{33} and on the relative location r_a of the sharp boundary between the plasma and the vacuum. The compression ratio was held fixed at approximately the value 3.5 planned for the Garching experiment. Both positive and negative values of Δ_{33} turn out to be stabilizing. More surprising is the dependence of stability on the location of the sharp boundary. It is seen that there is less stabilization when r_a is small. This suggests that force-free currents in the zone of nearly pressureless plasma play a role in the effect that has been observed. The conclusion is that our flux constraint $\mu(s) \equiv 0$, coupled with helically symmetric modifications of the wall geometry, can stabilize the $m = 1, k = 0$ mode significantly when the sharp boundary radius r_a is big enough.

In response to various criticisms of our stability theory [2], we have made longer runs of the code on the CDC 7600 computer at the Lawrence Berkeley

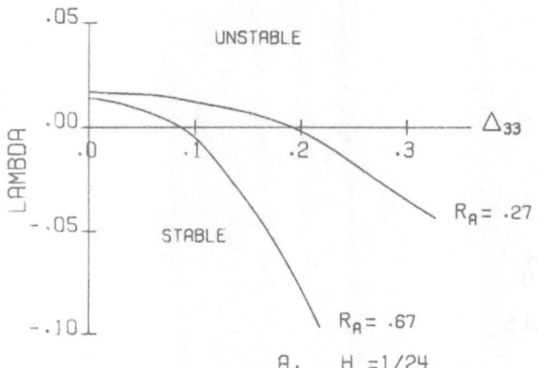

A.   H =1/24

Fig. 5.17 Effect of mesh and triangularity on growth rates.

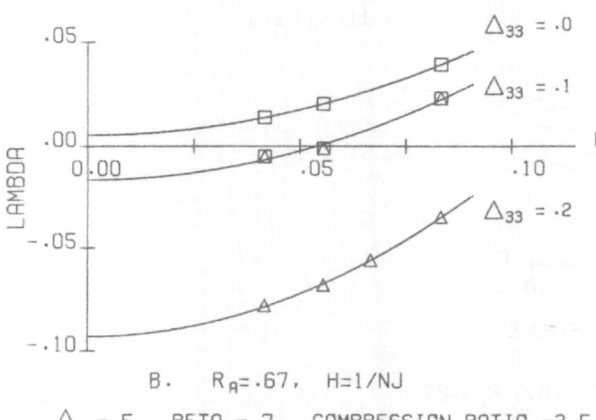

B. R_A=.67, H=1/NJ

$\triangle_1$ =.5, BETA =.7, COMPRESSION RATIO =3.5

Laboratory. The result about triangular cross sections continues to hold up on a finer mesh of $24 \times 48 \times 60$ points. It also holds up on a crude mesh of $16 \times 16 \times 192$ points modeling 12 adjacent periods of a helically symmetric equilibrium. For this model we have been able to confirm $m = 1$ stability for intermediate values $k = \frac{1}{2}, \frac{1}{3}, \frac{1}{6}$, and $\frac{1}{12}$ of the toroidal mode number. However, an excessively crude mesh size h_v in the toroidal direction has been found to give false predictions of $m = 1, k = 0$ stability in marginal cases. Finally, runs have been made with random shapes for the outer wall which suggest that only helically symmetric geometries are stabilizing.

In Section 2.6 we have already shown that the constraints in the plasma and vacuum regions are different. The magnetic field does not behave the same in the pressureless plasma region as it does in the vacuum region because its perturbations there are constrained to keep the rotational transform fixed. As we have seen, it may be possible to determine the rotational transform $\mu(s)$ so that the toroidal current $I(s)$ is zero, but this does not necessarily imply that

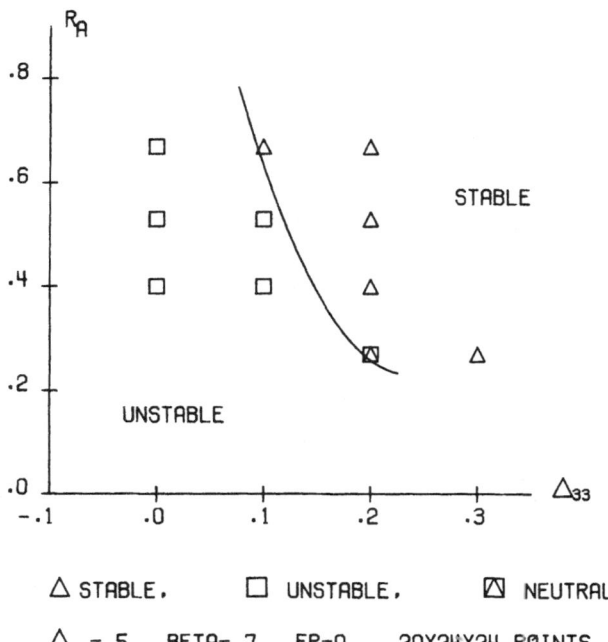

$\triangle$ STABLE.　　$\square$ UNSTABLE,　　◨ NEUTRAL

$\triangle_1 = .5$,　BETA$=.7$,　EP$=0$.,　20X24X24 POINTS

Fig. 5.18 Dependence of stability on triangularity and pressureless plasma.

the current density J vanishes. In particular, a straight helical $l = 1$ configuration with $\Delta_{33} = .1$ is likely to result in a dipole toroidal current density. This is, in fact, a possible explanation for the stability effect. Since $J = 0$ in the vacuum region, the force-free current disappears as the vacuum grows larger and, therefore, the $m = 1, k = 0$ mode becomes unstable.

Unfortunately, the effect of the pressureless plasma must disappear after a certain time, since resistivity destroys the flux constraints in cold, low-density plasma. The situation is not so dangerous, however, if we introduce a plasma column with a lower compression ratio of 3 or 4. In that case, triangular cross sections result in enhanced wall stabilization. Runs of the code indicate that this effect is appreciably stronger than the classical wall stabilization predicted by sharp boundary theory. Consequently, for a high β stellarator with low compression ratio and triangular cross sections, there is some prospect for containment on the time scale of 1 msec.

5.5 High β Tokamaks

We now turn our attention to axially symmetric equilibria and consider the effect of the shape of the cross section on the stability of free surface modes for a high β Tokamak. Let the aspect ratio $1/\varepsilon = 3$ and let the plasma radius $r_a = \frac{2}{3}$.

Given the wall shape, we define the rotational transform by

$$\mu(R) = \tfrac{1}{2} - \tfrac{1}{4}R^2$$

and the initial pressure profile by $p(R) = p_0(R) + p_1$, where

$$p_0(R) = .02[1 - R^2]^2$$

and p_1 is a constant corresponding to the pressure jump at the free boundary. We define the toroidal flux so that the pressure jump is balanced by a corresponding jump of the toroidal field at the interface. More precisely, for the straight cylinder approximation we put

$$\tfrac{1}{2}B_T^2 + p_1 = \tfrac{1}{2}$$

in the plasma region and $B_T = 1$ in the vacuum region. As a result, we can take the poloidal field in the vacuum to be small together with the toroidal surface current. They should be just big enough to compensate for the toroidal drift.

We now define a one-parameter family of equilibria depending on p_1. We ask that the smallest distance from the plasma boundary to the outer wall be

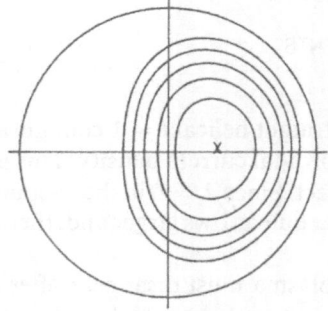

A. CIRCULAR CROSS SECTION

Fig. 5.19   Cross sections of two Tokamaks with critical $\beta = .18$.

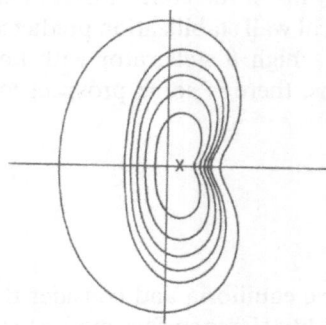

B. D-SHAPED CROSS SECTION

the same for all cases, and we investigate the dependence of stability on β for various cross sections.

For a circular cross section, the solution is shown in the first plot of Fig. 5.19. We can vary p_1 to arrive at stability results for the $m = 1$, $k = 1$ free surface mode. As we increase p_1, the value of β on the magnetic axis and the toroidal surface current both increase, so that the free surface mode becomes unstable. Thus we can determine the critical β for a given cross section. The second plot of Fig. 5.19 shows the flux surfaces for a cross section shaped like a doublet. This is obtained by setting $\Delta_{10} = -.2$, $\Delta_{20} = -.3$, and $\Delta_{30} = -.05$ in the wall formula

$$r_b = 1 + \Delta_{10} \cos U + \Delta_{20} \cos 2U + \Delta_{30} \cos 3U.$$

It turns out that in both cases the critical β is .18. To bring the plasma surface further away from the wall, a larger poloidal field is required, which in turn yields a lower critical β.

5.6 Discussion

The method we have described and the results we have presented suggest a number of open questions. It would be desirable to develop a two-dimensional version of the code in the helically symmetric case to test the stability theorem about triangular cross sections on a finer mesh than is feasible in three dimensions. One should investigate to what extent spectral methods, and especially fast Fourier transform, can be used to speed up the procedure and make it more accurate. A linear stability theory of our artifically time-dependent system of partial differential equations should be worked out. This would enable one to derive precise formulas relating the eigenvalues of our system to those of magnetohydrodynamics. The present version of the theory and the code that implements it need wider acceptance as a tool for the understanding and design of experiments. In this context, we cite the preliminary work on triangular cross sections now under way at Garching. Finally, to increase confidence in the physical relevance of the code, it would be desirable to make a more detailed comparison of the stabilizing effects of finite Larmor radius and finite mesh size.

References

1. Barnes, D. C. and Brackbill, J. U.: Computation of magnetohydrodynamic flow in a magnetically confined plasma. Nuc. Sci. Eng. **64**, 18–32 (1977).
2. Barnes, D., Brackbill, J., Dagazian, R., Freidberg, J., Schneider, W., and Betancourt, O.: Analytic and numerical studies of Scyllac equilibria. In *Plasma Physics and Controlled Nuclear Fusion Research. Sixth Conference Proceedings, Nuclear Fusion, Supplement, 1977, Vol. II.* IAEA, Vienna (1977), pp. 203–211.
3. Bauer, F., Garabedian, P., and Korn, D.: *Supercritical Wing Sections III*. Springer-Verlag, New York (1977).
4. Bernstein, I. B., Frieman, E. A., Kruskal, M. D., and Kulsrud, R. M.: An energy principle for hydromagnetic stability problems. Proc. Roy. Soc. London Ser. A **244**, 17–40 (1958).
5. Betancourt, O.: Three-dimensional computation of magnetohydrodynamic equilibrium of toroidal plasma without axial symmetry. ERDA Research and Development Report MF-67 and COO-3077-49, New York University (1974).
6. Betancourt, O. and Garabedian, P.: Computer simulation of the toroidal equilibrium and stability of a plasma in three dimensions. Proc. Natl. Acad. Sci. USA **72**, 926–927 (1975).
7. Betancourt, O. and Garabedian, P.: Equilibrium and stability code for a diffuse plasma. Proc. Natl. Acad. Sci. USA **73**, 984–987 (1976).
8. Betancourt, O. and Garabedian, P.: A method of computational magnetohydrodynamics defining stable Scyllac equilibria. Proc. Natl. Acad. Sci. USA **74**, 393–397 (1977).
9. Blank, A., Grad, H., and Weitzner, H.: Toroidal high-β equilibria. In *Plasma Physics and Controlled Nuclear Fusion Research, Third Conference Proceedings*. IAEA, Vienna (1969), pp. 607–617.
10. Bloch, E.: Numerical solution of free boundary problems by the method of steepest descent. Phys. Fluids (Suppl. II) **12**, 129–132 (1967).
11. Brackbill, J. U.: Numerical magnetohydrodynamics for high-beta plasmas. In Alder, B. and Fernbach, S. (eds): *Controlled Fusion*. Academic, New York (1976), pp. 1–41.
12. Chodura, R. and Schlüter, A.: In search of stable 3-D MHD-equilibria. *Second European Conference on Computational Physics*. Max Planck Institut für Plasma Physik, Garching (1976).
13. Freidberg, J. P. and Grossmann, W.: Magnetohydrodynamic stability of a sharp boundary model of tokamak. Phys. Fluids **18**, 1494–1506 (1975).
14. Friedman, N.: Computation of the equilibrium of a plasma with helical symmetry. AEC Computing and Applied Mathematics Center, New York University, Report NYO-1480-161 (1971).
15. Fünfer, E., Kaufmann, M., Lotz, W., Neuhauser, J., Schramm, G., and Seidel, U.: High-beta stellarator experiments on Isar Tl. Nuclear Fusion **15**, 133–142 (1975).
16. Fünfer, E., Kaufmann, M., Neuhauser, J., and Schramm, G.: Recent high-β stellarator experiments at Garching. In *The Seventh European Conference on Controlled Fusion and Plasma Physics*, Vol. II. Centre de Researches en Physique des Plasmas, Lausanne, Switzerland (1976), pp. 151–160.
17. Garabedian, P.: *Partial Differential Equations*. Wiley, New York (1964).
18. Garabedian, P.: Estimation of the relaxation factor for small mesh size. Math. Tables Aids Comput. **10**, 183–185 (1956).
19. Garabedian, P. and Schiffer, M.: Convexity of domain functionals. J. Analyse Math. **2**, 281–368 (1953).

20. Grad, H.: Toroidal containment of a plasma. Phys. Fluids 10, 137–154 (1967).
21. Grad, H., Hu, P. N., Stevens, D. C., and Turkel, E.: Classical plasma diffusion. In *Plasma Physics and Controlled Nuclear Fusion Research*, Nuclear Fusion, Supplement, *Vol. II*. IAEA, Vienna (1977), pp. 355–365.
22. Grad, H. and Rubin, H.: Hydromagnetic equilibrium and force-free fields. Proc. 2nd UN Int. Conf. Peaceful Uses Atomic Energy (United Nations, Geneva) 31, 190–197 (1958).
23. Grimm, R. C., Greene, J. M., and Johnson, J. L.: Computation of the magnetohydrodynamic spectrum in axisymmetric toroidal confinement systems. Alder, B. and Fernbach, S. (eds): *Controlled Fusion*. Academic, New York (1976), pp. 253–280.
24. Herrnegger, F., Kaufmann, M., Lortz, D., Neuhauser, J., Nührenberg, J., Schneider, W., and Schramm, G.: Experimental and theoretical study of the FLR stabilization in the high-beta stellarator. Equilibrium and stability of MHD-configurations with zero rotational transform. Plasma Phys. Contr. Nucl. Fusion Res. 2, 183–191 (1976).
25. Herrnegger, F. and Schneider, W.: Ideal MHD stability of $m \geq 2$ modes in diffuse high-β, $l = 1$ equilibria. Nucl. Fusion 16, 925–936 (1976).
26. Jameson, A.: Accelerated iteration schemes for transonic flow calculations using fast Poisson solvers. ERDA Research and Development Report COO-3077-82, New York University (1975).
27. Kappraff, J. M.: Bifurcated stability of a family of stellarator equilibria. Phys. Fluids 19, 675–682 (1976).
28. Kruskal, M. D. and Kulsrud, R. M.: Equilibrium of a magnetically confined plasma in a toroid. Phys. Fluids 1, 265–274 (1958).
29. Lax, P. D. and Wendroff, B.: Systems of conservation laws. Commun. Pure Appl. Math. 13, 217–237 (1960).
30. Lüst, R., Richtmyer, R. D., Rotenberg, A., Suydam, B. R., and Levy, D.: The hydromagnetic stability of a toroidal gas discharge. AEC Computing and Applied Mathematics Center, New York University, Report NYO-2882 (1960).
31. Lundquist, S.: On the stability of magnetohydrostatic fields. Phys. Rev. 83, 307–311 (1951).
32. Moser, J.: Lectures on Hamiltonian systems. Mem. AMS 81, 1–60 (1968).
33. Potter, D.: Waterbag methods in magnetohydrodynamics. In Alder, B. and Fernbach, S. (eds): *Controlled Fusion*. Academic, New York (1976), pp. 43–83.
34. Ribe, F.: Free-boundary solutions for high-beta stellarators of large aspect ratio. Tech. Rep. LA-4098 (Los Alamos Scientific Laboratory) (1969).
35. Shafranov, V. D.: Hydromagnetic stability of a current-carrying pinch in a strong longitudinal magnetic field. Sov. Phys. Tech. Phys. 15, 175–183 (1970).
36. Shohet, J. L., Anderson, D. T., Betancourt, O. L., and Tataronis, J. A.: Three-dimensional magnetohydrodynamic equilibria of toroidal stellarators. Phys. Rev. Lett. 35, 1433–1436 (1975).
37. Weitzner, H.: Free boundary long helical wavelength equilibria. Phys. Fluids 14, 658–670 (1971).

Listing of the Code with Comment Cards

```
1. OUTPUT FROM A SAMPLE RUN


INPUT DATA AS READ FROM INPUT DATA DECK


       EP   RBOUND     QLZ     NRUN
   0.0000    .5000  9.4248        2


     DELO     DEL1     DEL2     DEL3    DEL10    DEL20    DEL30
   0.0000   0.0000   0.0000   0.0000   0.0000   0.0000   0.0000


    DEL22    DEL33
   0.0000   0.0000


      XPR       PO     AMUO     AMU1     AMU2
   1.0000    .2840   1.6000   0.0000   0.0000


      ALF      AMP
   0.0000    .0300


   FUR(1)   FUR(2)   FUR(3)   FUR(4)   FUR(5)   FUR(6)   FUR(7)
   0.0000   0.0000   1.0000   0.0000   0.0000   0.0000   0.0000


   FUZ(1)   FUZ(2)   FUZ(3)   FUZ(4)   FUZ(5)   FUZ(6)   FUZ(7)
   0.0000   1.0000   0.0000   0.0000   0.0000   0.0000   0.0000


   FUX(1)   FUX(2)   FUX(3)   FUX(4)   FUX(5)   FUX(6)   FUX(7)
   0.0000   0.0000   0.0000   0.0000   0.0000   0.0000   0.0000


       NI       NJ       NK     ASYE      ERR
        8       14       14  .100E-03  .100E-06


      SA1      SA2      SA3       DT
   1.0000   1.0000    .1000    .0300


      SE1      SE2      SE3
  20.0000  20.0000   4.0000
```

```
   SAFI        NE      NVAC
 2.0000        50         1


    NR         NZ        NT       NAC
     3          2         5       100


    IC       TLIM
     8     800.00


   NIV         NV        NP       OM      SAFV       SE4       FV1       FV2
     8          2         1   1.8000    2.0000    1.0000    1.7420    2.3560


 PRINT1     PRINT2     PRINT3     PRINT4     PRINT5     PRINT6     PRINT7
 PSI ERR    BOU ERR   DEL ENER  RMA CO(V)  JAC RATIO


 PRINT8


   NRA1       NRA2      NZA1      NZA2       MK1       MK2       MK3       MK4
     3          5         2         4        12        11        22        21
```

ITER	PSI ERR	BOU ERR	DEL ENER	RMA CO(V)	JAC RATIO
1	.2656E+00	.4754E-01	.1894E+02	.3000E-01	.1064E+01
9	.2897E-01	.9043E-02	-.3067E-02	.3000E-01	.1015E+01
17	.1931E-01	.3022E-02	-.1360E-03	.3003E-01	.1006E+01
25	.2306E-01	.2467E-02	-.1003E-03	.3017E-01	.1008E+01
33	.2385E-01	.2474E-02	-.9585E-04	.3033E-01	.1013E+01
41	.2440E-01	.2508E-02	-.9539E-04	.3051E-01	.1018E+01
49	.2481E-01	.2542E-02	-.9751E-04	.3069E-01	.1023E+01
57	.2513E-01	.2575E-02	-.1006E-03	.3088E-01	.1028E+01
65	.2539E-01	.2608E-02	-.1040E-03	.3109E-01	.1033E+01
73	.2562E-01	.2643E-02	-.1077E-03	.3130E-01	.1039E+01
81	.2583E-01	.2680E-02	-.1115E-03	.3152E-01	.1044E+01
89	.2605E-01	.2717E-02	-.1155E-03	.3176E-01	.1049E+01
97	.2626E-01	.2755E-02	-.1197E-03	.3200E-01	.1055E+01
105	.2645E-01	.2801E-02	-.1241E-03	.3225E-01	.1060E+01
113	.2632E-01	.2917E-02	-.1311E-03	.3253E-01	.1066E+01
121	.2566E-01	.3134E-02	-.1426E-03	.3284E-01	.1073E+01
129	.2441E-01	.3454E-02	-.1592E-03	.3320E-01	.1080E+01
137	.2246E-01	.3898E-02	-.1826E-03	.3364E-01	.1087E+01
145	.1957E-01	.4508E-02	-.2156E-03	.3419E-01	.1096E+01
153	.1547E-01	.5338E-02	-.2619E-03	.3493E-01	.1105E+01
161	.1067E-01	.6271E-02	-.3226E-03	.3592E-01	.1116E+01
169	.7057E-02	.7069E-02	-.3890E-03	.3720E-01	.1128E+01
177	.4800E-02	.7744E-02	-.4589E-03	.3880E-01	.1141E+01
185	.4039E-02	.8375E-02	-.5335E-03	.4072E-01	.1154E+01
193	.3939E-02	.9034E-02	-.6151E-03	.4295E-01	.1168E+01
201	.3539E-02	.9832E-02	-.7117E-03	.4550E-01	.1183E+01
209	.2521E-02	.1078E-01	-.8323E-03	.4845E-01	.1199E+01
217	.1751E-02	.1175E-01	-.9760E-03	.5187E-01	.1217E+01
225	.1569E-02	.1271E-01	-.1139E-02	.5579E-01	.1235E+01
233	.1765E-02	.1375E-01	-.1324E-02	.6026E-01	.1254E+01
241	.1924E-02	.1488E-01	-.1539E-02	.6528E-01	.1274E+01
249	.2041E-02	.1612E-01	-.1786E-02	.7084E-01	.1295E+01
257	.2235E-02	.1753E-01	-.2078E-02	.7698E-01	.1317E+01
265	.2338E-02	.1908E-01	-.2422E-02	.8376E-01	.1340E+01
273	.2456E-02	.2074E-01	-.2823E-02	.9123E-01	.1366E+01
281	.2530E-02	.2254E-01	-.3287E-02	.9944E-01	.1393E+01
289	.2786E-02	.2449E-01	-.3824E-02	.1085E+00	.1421E+01
297	.3464E-02	.2659E-01	-.4444E-02	.1183E+00	.1453E+01
305	.4456E-02	.2885E-01	-.5156E-02	.1289E+00	.1486E+01
313	.5456E-02	.3127E-01	-.5971E-02	.1405E+00	.1523E+01
321	.6305E-02	.3386E-01	-.6903E-02	.1529E+00	.1564E+01
329	.7187E-02	.3660E-01	-.7955E-02	.1663E+00	.1608E+01
337	.8141E-02	.3952E-01	-.9137E-02	.1808E+00	.1655E+01
345	.9201E-02	.4261E-01	-.1046E-01	.1964E+00	.1707E+01
353	.1024E-01	.4586E-01	-.1193E-01	.2130E+00	.1764E+01
361	.1118E-01	.4927E-01	-.1354E-01	.2308E+00	.1826E+01

3D RESULTS

PLASMA ENERGY= .595858235E+01

VACUUM ENERGY= .128780422E+02

TOTAL ENERGY = .188365741E+02

FOURIER COEFFICIENTS MAGNETIC AXIS

	CONST	SIN(V)	COS(V)	SIN(2V)	COS(2V)
R	.00003	.00053	.23076	-.00001	.00001
Z	-.00002	.23082	-.00050	-.00000	.00000

FOURIER COEFFICIENTS VACUUM AXIS

	CONST	SIN(V)	COS(V)	SIN(2V)	COS(2V)
R	.00002	.00031	.13578	-.00001	.00001
Z	-.00001	.13580	-.00030	-.00000	-.00000

FOURIER COEFFICIENTS FREE BOUNDARY

	CONST	SIN(V)	COS(V)	SIN(2V)	COS(2V)
CONST	.49454	.00001	-.00002	-.00000	.00000
SIN(U)	-.00002	.14721	-.00040	-.00001	-.00000
COS(U)	.00002	.00042	.14717	-.00001	.00000
SIN(2U)	.00000	.00003	.00001	.00309	.00003
COS(2U)	-.00000	-.00000	.00001	-.00004	.00309
SIN(3U)	.00000	.00001	.00001	-.00000	-.00000
COS(3U)	-.00000	-.00000	.00000	-.00000	.00000

PLASMA REGION

RADIUS	BTOR	BPOL	TCURR	PCURR	MU	P	BETA
0.000	1.030	0.000	0.000	9.813	1.600	.291	.356
.132	1.032	.157	.134	9.822	1.600	.270	.333
.226	1.035	.257	.379	9.839	1.600	.229	.289
.294	1.038	.328	.624	9.854	1.600	.188	.244
.349	1.040	.386	.869	9.867	1.600	.147	.195
.396	1.041	.436	1.113	9.878	1.600	.105	.144
.438	1.043	.482	1.358	9.888	1.600	.063	.089
.476	1.044	.524	1.603	9.898	1.600	.021	.031

VACUUM REGION

TOROIDAL CURRENT= 1.548   POLOIDAL CURRENT= 9.788

RADIUS	BTOR	BPOL
.531	1.034	.474
.603	1.035	.414
.675	1.036	.367
.747	1.037	.331
.819	1.038	.301
.892	1.038	.276
.964	1.038	.256

	DESCENT COEFFICIENTS			RELAXATION FACTORS	
TIME	E1/A1	E2/A2	E3/A3	OM1	OM2
0.00	20.000	20.000	40.000	1.800	1.800
.24	20.000	20.000	40.000	1.800	1.800
.48	20.000	20.000	40.000	1.800	1.800
.72	20.000	20.000	40.000	1.800	1.800
.96	20.000	20.000	40.000	1.800	1.800
1.20	20.000	20.000	40.000	1.800	1.800
1.44	20.000	20.000	40.000	1.800	1.800
1.68	20.000	20.000	40.000	1.800	1.800
1.92	20.000	20.000	40.000	1.800	1.800
2.16	20.000	20.000	40.000	1.800	1.800
2.40	20.000	20.000	40.000	1.800	1.800
2.64	20.000	20.000	40.000	1.800	1.800
2.88	20.000	20.000	40.000	1.800	1.800
3.12	19.543	19.543	39.087	1.795	1.795
3.36	18.031	18.031	36.063	1.771	1.771
3.60	15.860	15.860	31.721	1.732	1.718
3.84	13.287	13.287	26.575	1.696	1.625
4.08	10.564	10.476	20.940	1.661	1.526
4.32	7.920	7.630	15.208	1.629	1.450
4.56	5.602	5.020	10.077	1.601	1.390
4.80	4.300	3.364	7.021	1.588	1.350
5.04	3.620	2.355	5.338	1.594	1.333
5.28	3.276	1.754	4.469	1.601	1.347
5.52	3.053	1.428	4.057	1.608	1.388
5.76	2.740	1.149	3.783	1.614	1.423
6.00	2.353	.850	3.370	1.620	1.449
6.24	1.997	.641	2.965	1.623	1.467
6.48	1.752	.525	2.674	1.623	1.476
6.72	1.593	.505	2.462	1.621	1.477
6.96	1.495	.487	2.319	1.620	1.472
7.20	1.404	.466	2.230	1.618	1.464
7.44	1.278	.446	2.131	1.617	1.458
7.68	1.115	.431	2.000	1.616	1.454
7.92	.956	.463	1.877	1.616	1.453
8.16	.827	.457	1.768	1.615	1.454
8.40	.742	.449	1.678	1.615	1.457
8.64	.693	.484	1.611	1.615	1.461
8.88	.643	.574	1.565	1.614	1.466
9.12	.576	.694	1.518	1.614	1.470
9.36	.497	.809	1.463	1.613	1.474
9.60	.417	.928	1.408	1.612	1.479
9.84	.347	1.023	1.355	1.612	1.484
10.08	.296	1.109	1.307	1.611	1.489
10.32	.270	1.177	1.266	1.610	1.496

2. FORTRAN LISTING

```
      PROGRAM BETA(INPUT=65,OUTPUT=514,TAPE3=514,TAPE4=514)
C     CALLS OTHER SUBROUTINES,CONTROLS INPUT-OUTPUT,COMPUTES DESCENT
C     COEFFICIENTS AND FOURIER COEFFICIENTS OF THE SOLUTION
      COMMON /PRINT/ IX(8),IJ(8),JX(50),PJX(50),NNJ,NJX
      COMMON RO(10,30,30),AL(10,30,30),XO(10,30,30),XL(10,30,30),R(30,30
     1),Z(30,30),RU(30,30),ZU(30,30),RV(30,30),ZV(30,30),X(30,30),KA(30)
     2,ZA(30),RN(30),ZN(30),RB1(30),RB2(30),ZB1(30),ZB2(30),RBU1(30),RBU
     32(30),ZBU1(30),ZBU2(30),RBV1(30),RBV2(30),ZBV1(30),ZBV2(30),HB1(30
     4),HB2(30)
      COMMON /AUX/ RR2,ZZ2,RRA,ZZA,XX(30),E1(10,30),E2(10,30),F1(10,30),
     1F2(10,30),G1(10,30),G2(10,30),P1(10,30),P2(10,30),Q1(10,30),Q2(10,
     230),V1(10,30),V2(10,30),U1(10,30),U2(10,30),UV1(10,30),UV2(10,30),
     3D(10,30),DD(10,30),BK(10,30),CA1(30),CA2(30),CB1(30),CB2(30),CC1(3
     40),CC2(30),CD1,CD2,CE1(30),CE2(30),CF1(30),CF2(30),CG1(30),CG2(30)
     5,CL1(30),CL2(30),CM1(30),CM2(30),CN1(30),CN2(30),XA1(30),XA2(30),X
     6B1(30),XB2(30),XC1(30),XC2(30),XG1(30),XG2(30),XN1(30),XN2(30),XD1
     7(30),XD2(30),XF1(30),XF2(30),XP1(30),XP2(30),XQ1(30),XQ2(30),SPR(1
     80)
      COMMON /FOU/ SV(7,30),SU(7,30),SFI(7,30),SRO(7,7),XR(7),XZ(7),YR(7
     1),YZ(7)
      COMMON /PLOT/ NRA1,NRA2,NZA1,NZA2,MK1,MK2,MK3,MK4,M1,M2,M3,M4,K1,K
     12,K3,K4,RNAME(7),ZNAME(7)
      COMMON /POT/ RVA(30),ZVA(30),BPV(10),BTV(10),PT(10,30,30),PP(10,30
     1,30)
      COMMON /INP/ Q(10),QT(10),AM(10),PR(10),QQ(10),PC(10),TC(10),BPP(1
     10),BTP(10),BET(10)
      COMMON /AC/ EN(100,5),ET(5),AA1(5),AA2(5),AA3(5),AVE(5),NAC,NE,SAF
     1I,SAFV,SAFPSI,SAFRO,SAFAX
      COMMON /CPL/ NI,NJ,NK,EP,ZLE,GAM,SM,N1,N2,N3,N4,N5,NVAC,HS,HU,HV,P
     1I,RX,RY,E4,A1,A2,A3,A4,A5,A6,HU4,HV4,HUV,IC,IO,SA1,SA2,SA3,SE1,SE2
     2,SE3,DT,RA1,RA2,RA3,RE1,RE2,RE3,ENER,FAXIS,DG1,DG2,DG3,DH1,DH2,DH3
     3,NPLOT,I1
      COMMON /CVA/ NIV,FV1,FV2,NV,NP,OM,PM,HR,H1,H2,H3,C1,C2,N6,EVAC,ETO
     1T,A11,A22,A12,HR4,DTB,FAXV,OM1,OM2,PM1,PM2
      COMMON /FUNC/ ALF,RBOU,DELO,DEL1,DEL2,DEL3,DEL10,DEL20,DEL30,DEL22
     1,DEL33,PO,XPR,AMUO,AMU1,AMU2,AMP,FUR(7),FUZ(7),FUX(7),NRUN
      DATA VAR/0.0/
C     READ INPUT DATA
      READ 730
      READ 680, EP,RBOU,QLZ,NRUN
      RX=1.0
      RY=RBOU
      READ 730
      READ 690, DELO,DEL1,DEL2,DEL3,DEL10,DEL20,DEL30
      READ 730
      READ 690, DEL22,DEL33
      READ 730
      READ 700, XPR,PO,AMUO,AMU1,AMU2
```

```
      READ 730
      READ 710,  ALF,AMP
      READ 730
      READ 720,  (FUR(I),I=1,7)
      READ 730
      READ 720,  (FUZ(I),I=1,7)
      READ 730
      READ 720,  (FUX(I),I=1,7)
      READ 730
      READ 620,  NI,NJ,NK,ASYE,ERR
      READ 730
      READ 610,  SA1,SA2,SA3,DT
      READ 730
      READ 630,  SE1,SE2,SE3
      READ 730
      READ 640,  SAFI,NE,NVAC
      READ 730
      READ 650,  NR,NZ,NT,NAC
      READ 730
      READ 660,  IC,TLIM
      IF (NVAC.GT.0) GO TO 10
      NP=1
      GO TO 20
   10 READ 730
      READ 670,  NIV,NV,NP,OM,SAFV,SE4,FV1,FV2
   20 READ 730
      READ 730,  (IX(I),I=1,7)
      READ 730
      READ 730,  IX(8)
      READ 730
      READ 740,  NRA1,NRA2,NZA1,NZA2,MK1,MK2,MK3,MK4
C     PRINT INPUT DATA
      PRINT 750
      PRINT 760,  EP,RBOU,QLZ,NRUN
      PRINT 770,  DELO,DEL1,DEL2,DEL3,DEL10,DEL20,DEL30
      PRINT 780,  DEL22,DEL33
      PRINT 790,  XPR,PO,AMU0,AMU1,AMU2
      PRINT 800,  ALF,AMP
      PRINT 810,  (FUR(I),I=1,7)
      PRINT 820,  (FUZ(I),I=1,7)
      PRINT 830,  (FUX(I),I=1,7)
      PRINT 840,  NI,NJ,NK,ASYE,ERR
      PRINT 850,  SA1,SA2,SA3,DT
      PRINT 860,  SE1,SE2,SE3
      PRINT 870,  SAFI,NE,NVAC
      PRINT 880,  NR,NZ,NT,NAC
      PRINT 890,  IC,TLIM
      IF (NVAC.LT.0) GO TO 30
      PRINT 900,  NIV,NV,NP,OM,SAFV,SE4,FV1,FV2
   30 CONTINUE
      PRINT 910, (IX(I),I=1,7)
      PRINT 920,  IX(8)
      PRINT 930,  NRA1,NRA2,NZA1,NZA2,MK1,MK2,MK3,MK4
C     DEFINE CONSTANTS
      PI=3.1415926535898
      FAXIS=1.0
      FAXV=1.0
```

```
        ELIM=2.0
        SAFRO=SAFI*SA1
        SAFPSI=SAFI*SA2
        SAFAX=SAFI*SA3
        IT=0
        I1=NI/2
        IO=50*IC
        N1=NJ+1
        N2=NJ+2
        N3=NI-1
        N4=NK+1
        N5=NK+2
        NJA=5
        IF (NVAC.LT.0) NJA=3
        IF (EP.LT.0.00001) GO TO 40
        ZLE=(2.0*PI)/(EP*QLZ)
        GO TO 50
40      ZLE=QLZ
50      SM=1.0
        GAM=5.0/3.0
        HS=SM/N3
        NPLOT=-1
        HU=1.0/NJ
        HV=1.0/NK
        IF (NVAC.LT.0) GO TO 60
        N6=NIV-1
        HR=1.0/N6
        H1=1.0/(HR*HR)
        H2=1.0/(HU*HU)
        H3=1.0/(HR*HU)
        HR4=0.25*H1
        DTB=DT/SE4
        PM=1.0-OM
60      CONTINUE
        A1=EP/8.0
        A2=1.0/(4.0*HS)
        A3=1.0/(4.0*HU*HU*ZLE*ZLE)
        A4=1.0/(4.0*HV*HV*ZLE*ZLE)
        A5=1.0/(4.0*HU*HV*ZLE*ZLE)
        A6=1.0/(8.0*HS)
        HU4=1.0/(4.0*HU*HU)
        HV4=1.0/(4.0*HV*HV)
        HUV=1.0/(4.0*HU*HV)
        RA1=SA1/(DT*DT)
        RA2=SA2/(DT*DT)
        RE1=SE1/DT
        RE2=SE2/DT
        RA3=SA3/(DT*DT)
        RE3=SE3/DT
        E4=EP/4.0
        NE1=NE-1
C       DEFINE MATRICES FOR FOURIER ANALYSIS
        DO 70 K=1,N5
        V=2.0*PI*(K-2)*HV
        SV(1,K)=1.0
        SV(2,K)=SIN(V)
        SV(3,K)=COS(V)
```

```
         SV(4,K)=SIN(2.0*V)
         SV(5,K)=COS(2.0*V)
         SV(6,K)=SIN(3.0*V)
   70    SV(7,K)=COS(3.0*V)
         DO 80 J=1,N2
         U=2.0*PI*(J-2)*HU
         SU(1,J)=1.0
         SU(2,J)=SIN(U)
         SU(3,J)=COS(U)
         SU(4,J)=SIN(2.0*U)
         SU(6,J)=SIN(3.0*U)
         SU(7,J)=COS(3.0*U)
   80    SU(5,J)=COS(2.0*U)
         ITER=0
         NTER=1000
         KTER=1000
C        COMPUTES AXIALLY SYMMETRIC SOLUTION USED TO INITIALIZE 3-D PROGRAM
         CALL ASYM (ASYE)
C        COMPUTES OUTER WALL SHAPE AND INITIALIZES 3-D SOLUTION
         CALL SURF
C        CHOOSES DATA TO BE PRINTED ACCORDING TO INPUT
         CALL PRNT
C        SETS INITIAL TIME DERIVATIVE FOR R AND PSI EQUAL TO ZERO
         DO 100 K=1,N5
         DO 90 J=1,N2
         DO 90 I=1,NI
         XO(I,J,K)=RO(I,J,K)
   90    XL(I,J,K)=AL(I,J,K)
  100    CONTINUE
         RIN=0.0
         ZIN=0.0
C        COMPUTATION OF AXIS FOURIER COEFFICIENTS WHICH WILL BE FIXED FOR
C        NT ITERATIONS
         DO 110 K=2,N4
         RIN=RIN+RA(K)*SV(NR,K)
  110    ZIN=ZIN+ZA(K)*SV(NZ,K)
         RIN=RIN*2.0*HV
         ZIN=ZIN*2.0*HV
         IF (NR.EQ.1) RIN=0.5*RIN
         IF (NZ.EQ.1) ZIN=0.5*ZIN
         IF (NVAC.LT.0) GO TO 130
         RIN1=0.0
         ZIN1=0.0
         DO 120 K=2,N4
         RIN1=RIN1+RVA(K)*SV(NR,K)
  120    ZIN1=ZIN1+ZVA(K)*SV(NZ,K)
         RIN1=RIN1*2.0*HV
         ZIN1=ZIN1*2.0*HV
         IF (NR.EQ.1) RIN1=0.5*RIN1
         IF (NZ.EQ.1) ZIN1=0.5*ZIN1
  130    CONTINUE
C        INITIAL VALUES FOR DESCENT COEFFICIENTS
         DO 140 I=1,NE
         EN(I,1)=SE1/SAFRO
         EN(I,2)=SE2/SAFPSI
  140    EN(I,3)=SE3/SAFAX
         IF (NVAC.LT.0) GO TO 160
```

```
       DO 150 I=1,NE
       EN(I,4)=(2.0-OM)/(OM*DT*SAFV)
  150 EN(I,5)=EN(I,4)
       OM1=OM
       OM2=OM
       PM1=1.0-OM1
       PM2=1.0-OM2
  160 CONTINUE
       DT2=2.0*DT
       RA1=SA1/(DT*DT)
       RA2=SA2/(DT*DT)
       RA3=SA3/(DT*DT)
       RE1=SE1/DT
       RE2=SE2/DT
       RE3=SE3/DT
       ETOT1=0.0
       EVNE=0.0
       ERBO=0.0
       ERBO1=0.0
       ERVA=0.0
       INI=0
       EA1=SE1
       EA2=SE2
       EA3=SE3
       DO 170 J=1,5
  170 AA2(J)=0.0
       REWIND 3
       REWIND 4
C      SETS INITIAL TIME DERIVATIVE FOR MAGNETIC AXIS EQUAL TO ZERO
       DO 180 K=1,N5
       RN(K)=RA(K)
  180 ZN(K)=ZA(K)
  190 CONTINUE
       IF (NVAC.LT.0) GO TO 200
C      COMPUTES NV ITERATIONS OF VACUUM EQUATIONS
       CALL TSOR (ERVA,EVNE,IT)
  200 CONTINUE
       NCO=0
  210 CONTINUE
       DG1=1.0/(RA1+RE1)
       DG2=1.0/(RA2+RE2)
       DG3=1.0/(RA3+RE3)
       DH1=2.0*RA1+RE1
       DH2=2.0*RA2+RE2
       DH3=2.0*RA3+RE3
C      COMPUTES ONE ITERATION OF PLASMA EQUATIONS
       CALL TGRAD (BJA,SJA,EAX,ERO,EAL)
       NT1=NT+10
       IF (ITER.GT.NT1) GO TO 240
C      FIXES GIVEN AXIS FOURIER COEFFICIENTS
       ALF=(1.0+NT1-ITER)/10.0
       IF (ALF.GT.1.0) ALF=1.0
       SUM1=0.0
       SUM2=0.0
       DO 220 K=2,N4
       SUM1=SUM1+RA(K)*SV(NR,K)
  220 SUM2=SUM2+ZA(K)*SV(NZ,K)
```

```
      SUM1=SUM1*2.0*HV
      SUM2=SUM2*2.0*HV
      IF (NR.EQ.1) SUM1=0.5*SUM1
      IF (NZ.EQ.1) SUM2=0.5*SUM2
      DO 230 K=1,N5
      RA(K)=RA(K)+ALF*SV(NR,K)*(RIN-SUM1)
  230 ZA(K)=ZA(K)+ALF*SV(NZ,K)*(ZIN-SUM2)
  240 CONTINUE
      IF (NVAC.LT.0) GO TO 270
      IF (ITER.GT.NT1) GO TO 270
      SUM1=0.0
      SUM2=0.0
      DO 250 K=2,N4
      SUM1=SUM1+RVA(K)*SV(NR,K)
  250 SUM2=SUM2+ZVA(K)*SV(NZ,K)
      SUM1=SUM1*2.0*HV
      SUM2=SUM2*2.0*HV
      IF (NP.EQ.1) SUM1=0.5*SUM1
      IF (NZ.EQ.1) SUM2=0.5*SUM2
      DO 260 K=1,N5
      RVA(K)=RVA(K)+ALF*SV(NR,K)*(RIN1-SUM1)
  260 ZVA(K)=ZVA(K)+ALF*SV(NZ,K)*(ZIN1-SUM2)
  270 CONTINUE
      RATIO=SJA/BJA
      IF (RATIO.GE.40.0) GO TO 590
      NCO=NCO+1
      IF (NCO.LT.NP) GO TO 210
      IF (NVAC.LT.0) GO TO 280
C     COMPUTES ONE ITERATION OF FREE BOUNDARY EQUATION
      CALL TBO (ERBO,ERBO1)
  280 CONTINUE
      ITER=ITER+1
      IF (MOD(ITER,200).EQ.0) CALL CHEKPTX (VAR)
      NTER=NTER+1
      KTER=KTER+1
      IF (ITER.LT.NAC) GO TO 320
C     NEW VALUES FOR DESCENT COEFFICIENTS
      DO 300 J=1,NJA
      ET(J)=(AA1(J)-AA3(J))/DT2
      DO 290 I=1,NE1
  290 EN(I,J)=EN(I+1,J)
  300 EN(NE,J)=ABS(ET(J))/AA2(J)
      DO 310 J=1,NJA
      Y1=(EN(NE,J)-EN(NE-1,J))/(DT*0.5*(EN(NE,J)+EN(NE-1,J)))
      Y2=ABS(Y1)
      IF (Y2.GT.ELIM) EN(NE,J)=EN(NE-1,J)*(1.0+SIGN(ELIM,Y1)*DT)
  310 CONTINUE
  320 CONTINUE
      DO 330 J=1,NJA
      AA3(J)=AA2(J)
  330 AA2(J)=AA1(J)
      IF (ITER.LT.NAC) GO TO 360
      DO 350 J=1,NJA
      AVE(J)=0.0
      DO 340 I=1,NE
  340 AVE(J)=AVE(J)+EN(I,J)
  350 AVE(J)=AVE(J)/NE
```

```
          EA1=SAFRO*AVE(1)
          EA2=SAFPSI*AVE(2)
          EA3=SAFAX*AVE(3)
          RE1=EA1/DT
          RE2=EA2/DT
          RE3=EA3/DT
          IF (NVAC.LT.0) GO TO 360
          OM1=2.0/(1.0+SAFV*AVE(4)*DT)
          OM2=2.0/(1.0+SAFV*AVE(5)*DT)
          PM1=1.0-OM1
          PM2=1.0-OM2
      360 CONTINUE
C         PRINTOUT, TIME AND ERROR CRITERIA
          IF (KTER.GE.IO) GO TO 400
      370 IF (NTER.GE.IC) GO TO 410
      380 TMAX=AMAX1(EAX,ERO,EAL,ERBO)
          IF (TIM1.LT.TLIM) GO TO 390
          CALL CHEKPTX (VAR)
          CALL SECOND (TIM1)
          IF (TIM1.GE.TLIM) GO TO 590
      390 CONTINUE
          IF (TMAX.GT.ERR) GO TO 190
          GO TO 590
      400 KTER=0
          INI=0
          GO TO 370
      410 NTER=0
          ETOT=ENER+EVNE
          GEN=ETOT-ETOT1
          ETOT1=ETOT
          EVAC=EVNE
C         HEADING FOR ITERATION DATA PRINTED OUT
          IF (INI.EQ.0) PRINT 940, (JX(IJ(J)),J=1,NNJ)
          IF (INI.EQ.0) PRINT 950
          INI=1
          CALL SECOND (TIM1)
C         COMPUTATION OF FOURIER COEFFICIENTS
          DO 430 I=1,7
          XR(I)=0.0
          XZ(I)=0.0
          DO 420 K=2,N4
          XR(I)=XR(I)+RA(K)*SV(I,K)
      420 XZ(I)=XZ(I)+ZA(K)*SV(I,K)
          XR(I)=2.0*HV*XR(I)
      430 XZ(I)=2.0*HV*XZ(I)
          XR(1)=0.5*XR(1)
          XZ(1)=0.5*XZ(1)
          WRITE (3) ETOT,ERO,EAL,EAX
          WRITE (3) (XR(I),I=1,7),(XZ(I),I=1,7)
          IF (NVAC.LT.0) GO TO 460
          DO 450 I=1,7
          YR(I)=0.0
          YZ(I)=0.0
          DO 440 K=2,N4
          YR(I)=YR(I)+RVA(K)*SV(I,K)
      440 YZ(I)=YZ(I)+ZVA(K)*SV(I,K)
          YR(I)=2.0*HV*YR(I)
```

```
450 YZ(I)=2.0*HV*YZ(I)
    YR(1)=0.5*YR(1)
    YZ(1)=0.5*YZ(1)
    WRITE (4) ERBO,ERBO1,ERVA
    WRITE (4) (YR(I),I=1,7),(YZ(1),I=1,7)
460 CONTINUE
    IF (NVAC.GT.0) GO TO 500
    DO 490 K=2,N4
    DO 480 L=1,7
    SFI(L,K)=0.0
    DO 470 J=2,N1
470 SFI(L,K)=SFI(L,K)+RO(I1,J,K)*SU(L,J)
480 SFI(L,K)=SFI(L,K)*2.0*HU
490 SFI(1,K)=0.5*SFI(1,K)
    GO TO 540
500 CONTINUE
    DO 530 K=2,N4
    V=(K-2)*HV
    DO 520 L=1,7
    SFI(L,K)=0.0
    DO 510 J=2,N1
510 SFI(L,K)=SFI(L,K)+X(J,K)*SU(L,J)
520 SFI(L,K)=SFI(L,K)*2.0*HU
530 SFI(1,K)=0.5*SFI(1,K)
540 DO 570 L=1,7
    DO 560 M=1,7
    SRO(L,M)=0.0
    DO 550 K=2,N4
550 SRO(L,M)=SRO(L,M)+SFI(L,K)*SV(M,K)
560 SRO(L,M)=SRO(L,M)*2.0*HV
570 SRO(L,1)=0.5*SRO(L,1)
    Y00=SRO(1,1)
    Y10=SQRT(SRO(2,1)*SRO(2,1)+SRO(3,1)*SRO(3,1))
    Y20=SQRT(SRO(4,1)*SRO(4,1)+SRO(5,1)*SRO(5,1))
    Y30=SQRT(SRO(6,1)*SRO(6,1)+SRO(7,1)*SRO(7,1))
    Y01=SQRT(SRO(1,2)*SRO(1,2)+SRO(1,3)*SRO(1,3))
    Y11=SQRT(SRO(2,2)*SRO(2,2)+SRO(2,3)*SRO(2,3)+SRO(3,2)*SRO(3,2)+SRO
   1(3,3)*SRO(3,3))
    Y21=SQRT(SRO(4,2)*SRO(4,2)+SRO(4,3)*SRO(4,3)+SRO(5,2)*SRO(5,2)+SRO
   1(5,3)*SRO(5,3))
    Y31=SQRT(SRO(6,2)*SRO(6,2)+SRO(6,3)*SRO(6,3)+SRO(7,2)*SRO(7,2)+SRO
   1(7,3)*SRO(7,3))
    Y02=SQRT(SRO(1,4)*SRO(1,4)+SRO(1,5)*SRO(1,5))
    Y12=SQRT(SRO(2,4)*SRO(2,4)+SRO(3,4)*SRO(3,4)+SRO(2,5)*SRO(2,5)+SRO
   1(3,5)*SRO(3,5))
    Y22=SQRT(SRO(4,4)*SRO(4,4)+SRO(4,5)*SRO(4,5)+SRO(5,4)*SRO(5,4)+SRO
   1(5,5)*SRO(5,5))
    Y32=SQRT(SRO(6,4)*SRO(6,4)+SRO(7,4)*SRO(7,4)+SRO(6,5)*SRO(6,5)+SRO
   1(7,5)*SRO(7,5))
    Y03=SQRT(SRO(1,6)*SRO(1,6)+SRO(1,7)*SRO(1,7))
    Y13=SQRT(SRO(2,6)*SRO(2,6)+SRO(3,6)*SRO(3,6)+SRO(2,7)*SRO(2,7)+SRO
   1(3,7)*SRO(3,7))
    Y23=SQRT(SRO(4,6)*SRO(4,6)+SRO(5,6)*SRO(5,6)+SRO(4,7)*SRO(4,7)+SRO
   1(5,7)*SRO(5,7))
    Y33=SQRT(SRO(6,6)*SRO(6,6)+SRO(7,6)*SRO(7,6)+SRO(6,7)*SRO(6,7)+SRO
   1(7,7)*SRO(7,7))
    WRITE (3) Y00,Y10,Y20,Y30,Y01,Y11,Y21,Y31,Y02,Y12,Y22,Y32,Y03,Y13,
```

```
      1Y23,Y33
       Y1=EA1/SA1
       Y2=EA2/SA2
       Y3=EA3/SA3
       WRITE (3) Y1,Y2,Y3
       WRITE (4) OM1,OM2
C      SELECTION OF DATA TO BE PRINTED
       PJX(1)=EAX
       PJX(2)=ERO
       PJX(3)=EAL
       PJX(4)=ERBO
       PJX(5)=ERVA
       PJX(6)=GEN
       PJX(7)=RATIO
       PJX(8)=ERBO1
       DO 580 I=1,7
       PJX(I+8)=XR(I)
  580  PJX(I+15)=XZ(I)
       PJX(23)=Y00
       PJX(24)=Y10
       PJX(25)=Y20
       PJX(26)=Y30
       PJX(27)=Y01
       PJX(28)=Y11
       PJX(29)=Y21
       PJX(30)=Y31
       PJX(31)=Y02
       PJX(32)=Y12
       PJX(33)=Y22
       PJX(34)=Y32
       PJX(35)=Y03
       PJX(36)=Y13
       PJX(37)=Y23
       PJX(38)=Y33
       PJX(39)=EA1
       PJX(40)=EA2
       PJX(41)=EA3
       PJX(42)=OM1
       PJX(43)=OM2
       PRINT 960, ITER,(PJX(IJ(J)),J=1,NNJ)
       GO TO 380
  590  IF (RATIO.GE.30) PRINT 970, RATIO
       NPLOT=1
C      COMPUTE FINAL VALUES OF MAGNETIC FIELD AND CURRENTS IN PLASMA
C      REGION
       CALL TGRAD (BJA,SJA,EAX,ERO,EAL)
       IF (NVAC.LT.0) GO TO 600
C      COMPUTES FINAL VALUES OF MAGNETIC FIELD IN VACUUM REGION
       CALL TSOR (ERVA,EVNE,IT)
  600  CONTINUE
C      FINAL OUTPUT AND PLOTTING
       CALL FPRINT (ITER)
       CALL TPLOT (ITER)
       STOP
  610  FORMAT (4F10.5)
  620  FORMAT (3I5,2E10.3)
  630  FORMAT (3F10.5)
```

```
 640 FORMAT (F10.5,2I5)
 650 FORMAT (4I5)
 660 FORMAT (I5,F10.2)
 670 FORMAT (3I5,5F10.5)
 680 FORMAT (3F10.5,I5)
 690 FORMAT (8F10.5)
 700 FORMAT (5F10.5)
 710 FORMAT (2F10.5)
 720 FORMAT (7F10.5)
 730 FORMAT (8A10)
 740 FORMAT (8I5)
 750 FORMAT (1H1,9(/),6X39HINPUT DATA AS READ FROM INPUT DATA DECK//)
 760 FORMAT (13X2HEP,2X6HRBOUND,5X3HQLZ,4X4HNRUN/8X,3(F7.4,1X),2X1I5//)
 770 FORMAT (10X5HDELO 3X5HDEL1 3X5HDEL2 3X5HDEL3 3X5HDEL10,3X5HDEL20,3
    1X5HDEL30/8X7(F7.4,1X)///)
 780 FORMAT (10X5HDEL22,3X5HDEL33/8X2(F7.4,1X)///)
 790 FORMAT (12X3HXPR,6X2HPO,4X4HAMUO,4X4HAMU1,4X4HAMU2/8X5(F7.4,1X)///)
 800 FORMAT (12X3HALF,5X3HAMP/8X2(F7.4,1X)///)
 810 FORMAT (9X6HFUR(1),2X6HFUR(2),2X6HFUR(3),2X6HFUR(4),2X6HFUR(5),2X6
    1HFUR(6),2X6HFUR(7)/8X7(F7.4,1X)///)
 820 FORMAT (9X6HFUZ(1),2X6HFUZ(2),2X6HFUZ(3),2X6HFUZ(4),2X6HFUZ(5),2X6
    1HFUZ(6),2X6HFUZ(7)/8X7(F7.4,1X)///)
 830 FORMAT (9X6HFUX(1),2X6HFUX(2),2X6HFUX(3),2X6HFUX(4),2X6HFUX(5),2X6
    1HFUX(6),2X6HFUX(7)/8X7(F7.4,1X)///)
 840 FORMAT (13X2HNI,6X2HNJ,6X2HNK,6X4HASYE,7X3HERR/7X3(3X,I5),2(E10.3)
    1///)
 850 FORMAT (12X3HSA1,5X3HSA2,5X3HSA3,6X2HDT/8X3(F7.4,1X),F7.4//)
 860 FORMAT (12X3HSE1,5X3HSE2,5X3HSE3/8X3(F7.4,1X)//)
 870 FORMAT (1H1///11X4HSAFI,6X2HNE,4X4HNVAC/8XF7.4,2(3XI5)///)
 880 FORMAT (13X2HNR,6X2HNZ,6X2HNT,5X3HNAC/7X4(3XI5)///)
 890 FORMAT (13X2HIC,4X4HTLIM/10XI5,F8.2//)
 900 FORMAT (12X3HNIV,6X2HNV,6X2HNP,6X2HOM,4X4HSAFV,5X3HSE4,5X3HFV1,5X3
    1HFV2/7X,3(3XI5),5(1X,F7.4)///)
 910 FORMAT (9X,6HPRINT1,4X6HPRINT2,4X6HPRINT3,4X6HPRINT4,4X6HPRINT5,4X
    16HPRINT6,4X6HPRINT7/8X7(A10)///)
 920 FORMAT (9X,6HPRINT8/8XA10//)
 930 FORMAT (11X4HNRA1,4X4HNRA2,4X4HNZA1,4X4HNZA2,5X3HMK1,5X3HMK2,5X3HM
    1K3,5X3HMK4/7X8(3XI5)///)
 940 FORMAT (1H1///6X4HITER,5X8(A10,3X))
 950 FORMAT (1H0)
 960 FORMAT (5XI5,8E13.4)
 970 FORMAT (//6X19HRATIO EXCEEDS LIMIT,2XF6.2)
     END




     SUBROUTINE PRNT
C    CHOOSES DATA TO BE PRINTED ACCORDING TO INPUT
     COMMON /PRINT/ IX(8),IJ(8),JX(50),PJX(50),NNJ,NJX
     JX(1)=10HAXIS ERR
     JX(2)=10HRO ERR
     JX(3)=10HPSI ERR
     JX(4)=10HBOU ERR
     JX(5)=10HVAC ERR
     JX(6)=10HDEL ENER
```

```
      JX(7)=10HJAC RATIO
      JX(8)=10HVAC AXIS
      JX(9)=10HRMA CONS
      JX(10)=10HRMA SI(V)
      JX(11)=10HRMA CO(V)
      JX(12)=10HRMA SI(2V)
      JX(13)=10HRMA CO(2V)
      JX(14)=10HRMA SI(3V)
      JX(15)=10HRMA CO(3V)
      JX(16)=10HZMA CONS
      JX(17)=10HZMA SI(V)
      JX(18)=10HZMA CO(V)
      JX(19)=10HZMA SI(2V)
      JX(20)=10HZMA CO(2V)
      JX(21)=10HZMA SI(3V)
      JX(22)=10HZMA CO(3V)
      JX(23)=10HMK=00
      JX(24)=10HMK=10
      JX(25)=10HMK=20
      JX(26)=10HMK=30
      JX(27)=10HMK=01
      JX(28)=10HMK=11
      JX(29)=10HMK=21
      JX(30)=10HMK=31
      JX(31)=10HMK=02
      JX(32)=10HMK=12
      JX(33)=10HMK=22
      JX(34)=10HMK=32
      JX(35)=10HMK=03
      JX(36)=10HMK=13
      JX(37)=10HMK=23
      JX(38)=10HMK=33
      JX(39)=10HE1
      JX(40)=10HE2
      JX(41)=10HE3
      JX(42)=10HOM1
      JX(43)=10HOM2
      NNJ=0
      NJX=43
      DO 20 J=1,8
      DO 10 K=1,NJX
      IF (IX(J).NE.JX(K)) GO TO 10
      NNJ=NNJ+1
      IJ(NNJ)=K
   10 CONTINUE
   20 CONTINUE
      RETURN
      END



      SUBROUTINE TGRAD (BJA,SJA,EAX,ERO,EAL)
C     COMPUTES ONE ITERATION OF PLASMA EQUATIONS
      COMMON RO(10,30,30),AL(10,30,30),XO(10,30,30),XL(10,30,30),R(30,30
     1),Z(30,30),RU(30,30),ZU(30,30),RV(30,30),ZV(30,30),X(30,30),RA(30)
```

```
      2,ZA(30),RN(30),ZN(30),RB1(30),RB2(30),ZB1(30),ZB2(30),RBU1(30),RBU
      32(30),ZBU1(30),ZBU2(30),RBV1(30),RBV2(30),ZBV1(30),ZBV2(30),HB1(30
      4),HB2(30)
       COMMON /AUX/ RR2,ZZ2,RRA,ZZA,XX(30),E1(10,30),E2(10,30),F1(10,30),
      1F2(10,30),G1(10,30),G2(10,30),P1(10,30),P2(10,30),Q1(10,30),Q2(10,
      230),V1(10,30),V2(10,30),U1(10,30),U2(10,30),UV1(10,30),UV2(10,30),
      3D(10,30),DD(10,30),BK(10,30),CA1(30),CA2(30),CB1(30),CB2(30),CC1(3
      40),CC2(30),CD1,CD2,CE1(30),CE2(30),CF1(30),CF2(30),CG1(30),CG2(30)
      5,CL1(30),CL2(30),CM1(30),CM2(30),CN1(30),CN2(30),XA1(30),XA2(30),X
      6B1(30),XB2(30),XC1(30),XC2(30),XG1(30),XG2(30),XN1(30),XN2(30),XD1
      7(30),XD2(30),XF1(30),XF2(30),XP1(30),XP2(30),XQ1(30),XQ2(30),SPR(1
      80)
       COMMON /INP/ Q(10),QT(10),AM(10),PR(10),QQ(10),PC(10),TC(10),BPP(1
      10),BTP(10),BET(10)
       COMMON /AC/ EN(100,5),ET(5),AA1(5),AA2(5),AA3(5),AVE(5),NAC,NE,SAF
      1I,SAFV,SAFPSI,SAFRO,SAFAX
       COMMON /CPL/ NI,NJ,NK,EP,ZLE,GAM,SM,N1,N2,N3,N4,N5,NVAC,HS,HU,HV,P
      1I,RX,RY,E4,A1,A2,A3,A4,A5,A6,HU4,HV4,HUV,IC,IO,SA1,SA2,SA3,SE1,SE2
      2,SE3,DT,RA1,RA2,RA3,RE1,RE2,RE3,ENER,FAXIS,DG1,DG2,DG3,DH1,DH2,DH3
      3,NPLOT,I1
       COMMON /CVA/ NIV,FV1,FV2,NV,NP,OM,PM,HR,H1,H2,H3,C1,C2,N6,EVAC,ETO
      1T,A11,A22,A12,HR4,DTB,FAXV,OM1,OM2,PM1,PM2
       DO 10 J=1,3
   10 AA1(J)=0.0
       DO 20 I=1,NI
   20 SPR(I)=0.0
       ENER=0.0
       BJA=1000.0
       SJA=-1000.0
       CALL CBO (1)
       CALL CIN (1,BJA,SJA)
       BJA=1000.0
       SJA=-1000.0
       IF (NPLOT.LT.0) GO TO 40
       DO 30 I=1,NI
       BTP(I)=0.0
       BPP(I)=0.0
       BET(I)=0.0
       PC(I)=0.0
   30 TC(I)=0.0
       FAC1=HU*HV
       FAC=FAC1/ZLE
   40 CONTINUE
       ENER=0.0
       DO 50 I=1,NI
   50 SPR(I)=0.0
       EAX=0.0
       ERO=0.0
       EAL=0.0
       DO 160 K=2,N4
C     COEFFICIENT MATRICES ARE DENOTED BY 1 WHEN EVALUATED AT K+1/2 AND
C     BY 2 WHEN EVALUATED AT K-1/2
       CD2=CD1
       DO 60 J=1,N1
       RB2(J)=RB1(J)
       ZB2(J)=ZB1(J)
       RBU2(J)=RBU1(J)
```

```
       ZBU2(J)=ZBU1(J)
       RBV2(J)=RBV1(J)
       ZBV2(J)=ZBV1(J)
       HB2(J)=HB1(J)
       CA2(J)=CA1(J)
       CB2(J)=CB1(J)
       CC2(J)=CC1(J)
       CE2(J)=CE1(J)
       CF2(J)=CF1(J)
       CG2(J)=CG1(J)
       CL2(J)=CL1(J)
       CM2(J)=CM1(J)
       CN2(J)=CN1(J)
       XA2(J)=XA1(J)
       XB2(J)=XB1(J)
       XC2(J)=XC1(J)
       XG2(J)=XG1(J)
       XN2(J)=XN1(J)
       XD2(J)=XD1(J)
       XF2(J)=XF1(J)
       XP2(J)=XP1(J)
       XQ2(J)=XQ1(J)
       DO 60 I=1,NI
       U2(I,J)=U1(I,J)
       V2(I,J)=V1(I,J)
       UV2(I,J)=UV1(I,J)
       E2(I,J)=E1(I,J)
       F2(I,J)=F1(I,J)
       G2(I,J)=G1(I,J)
       P2(I,J)=P1(I,J)
    60 Q2(I,J)=Q1(I,J)
C      EVALUATES BOUNDARY QUANTITIES AT K+1/2
       CALL CBO (K)
C      EVALUATES PLASMA EQUATIONS COEFFICIENTS AT K+1/2
       CALL CIN (K,BJA,SJA)
       IF (NPLOT.LT.0) GO TO 80
C      COMPUTE VALUES OF PHYSICAL QUANTITIES FOR FINAL OUTPUT
       DO 70 J=2,N1
       E1(NI,J)=E1(NI,J)/2.0
       F1(NI,J)=F1(NI,J)/2.0
       G1(NI,J)=G1(NI,J)/2.0
       DO 70 I=1,N3
       PU=(AL(I,J+1,K)+AL(I+1,J+1,K)+AL(I,J+1,K+1)+AL(I+1,J+1,K+1)-AL(I,J
      1,K)-AL(I+1,J,K)-AL(I,J,K+1)-AL(I+1,J,K+1))/(4.0*HU)
       PV=(AL(I,J,K+1)+AL(I+1,J,K+1)+AL(I,J+1,K+1)+AL(I+1,J+1,K+1)-AL(I,J
      1,K)-AL(I+1,J,K)-AL(I,J+1,K)-AL(I+1,J+1,K))/(4.0*HV)
       BL=D(I,J)
       AK=0.5*(BK(I,J)+BK(I+1,J))
       BET(I)=BET(I)+(PR(I)*AK*FAC1)/P1(I,J)
       AJ=AK*BL
       X1=BK(I,J)/DD(I,J)
       X2=BK(I+1,J)/DD(I+1,J)
       Y1=0.5*(E1(I,J)*X1+E1(I+1,J)*X2)
       Y2=0.5*(G1(I,J)*X1+G1(I+1,J)*X2)
       X3=0.5*(F1(I,J)*X1+F1(I+1,J)*X2)
       X1=SQRT(Y1)*AJ
       X2=SQRT(Y2)*AJ
```

```
      X3=PV*Y1-PU*Y3
      X4=PV*Y3-PU*Y2
      FAC=FAC1/(ZLE*AJ)
      PC(I)=PC(I)+X4*FAC
      TC(I)=TC(I)+X3*FAC
      XO(I,J,K)=X3/(ZLE*X1)
      XL(I,J,K)=X4/(ZLE*X2)
      BPP(I)=BPP(I)+XO(I,J,K)*FAC1
   70 BTP(I)=BTP(I)+XL(I,J,K)*FAC1
      GO TO 160
   80 CONTINUE
C     SOLUTION OF PSI EQUATION AT MAGNETIC AXIS
      V=(K-2)*HV
      AL(1,2,K)=Q(1)*V
      SUM1=0.0
      DO 90 J=2,N1
   90 SUM1=SUM1+1.0/(G1(1,J)+G2(1,J))
      SUM1=-QT(1)/SUM1
      DO 100 J=2,N1
  100 AL(1,J+1,K)=AL(1,J,K)+SUM1/(G1(1,J)+G2(1,J))
      AL(1,1,K)=AL(1,N1,K)+QT(1)
C     ITERATION FOR PSI EQUATION
      DO 120 I=2,NI
      XL(I,1,K)=AL(I,1,K)
      DO 110 J=2,N1
      Y=AL(I,J,K)
      X1=(E1(I,J)+E1(I,J-1))*(AL(I,J,K+1)-Y)-(E2(I,J)+E2(I,J-1))*(Y-XL(I
     1,J,K-1))
      X2=(G1(I,J)+G2(I,J))*(AL(I,J+1,K)-Y)-(G1(I,J-1)+G2(I,J-1))*(Y-XL(I
     1,J-1,K))
      X3=F1(I,J)*(AL(I,J+1,K+1)-Y)+F1(I,J-1)*(Y-AL(I,J-1,K+1))-F2(I,J)*(
     1XL(I,J+1,K-1)-Y)-F2(I,J-1)*(Y-XL(I,J-1,K-1))
      DAL=-A4*X1-A3*X2+A5*X3
      AA=ABS(DAL)
      EAL=AMAX1(EAL,AA)
      PAS=DG2*(Y*DH2-DAL-RA2*XL(I,J,K))
      DAL=(PAS-AL(I,J,K))/DT
      AA1(2)=AA1(2)+DAL*DAL
      XL(I,J,K)=Y
  110 AL(I,J,K)=PAS
      AL(I,1,K)=AL(I,N1,K)+QT(I)
      AL(I,N2,K)=AL(I,2,K)-QT(I)
  120 XL(I,N2,K)=XL(I,2,K)-QT(I)
      HU2=2.0*HU
      HV2=2.0*HV
      HUV2=2.0*HU*HV
      HUU2=2.0*HU*HU
      HVV2=2.0*HV*HV
      W1=0.5*(RA(K)+RA(K+1))
      W2=0.5*(RA(K)+RN(K-1))
      SUMR=0.0
      SUMZ=0.0
C     ITERATION FOR R EQUATION
      DO 140 I=2,N3
      XO(I,1,K)=RO(I,1,K)
      X1=Q1(I,1)*(RB1(1)-W1)+Q2(I,1)*(RB2(1)-W2)
      X3=HB1(1)*(P1(I,1)-P1(I-1,1))+HB2(1)*(P2(I,1)-P2(I-1,1))
```

```
Y1=U1(I,1)
Y2=U2(I,1)
Y3=V1(I,1)
Y4=V2(I,1)
Y5=UV1(I,1)
Y6=UV2(I,1)
X5=(CA2(1)*Y4+CA1(1)*Y3)*(RO(I,2,K)-RO(I,1,K))
X7=CB1(1)*Y3+CE1(1)*Y1-2.0*CM1(1)*Y5+CB2(1)*Y4+CE2(1)*Y2-2.0*CM2(1
1)*Y6
X9=CC1(1)*Y3-CG1(1)*Y5+CC2(1)*Y4-CG2(1)*Y6
X11=CG1(1)*Y1-CC1(1)*Y5-CG2(1)*Y2+CC2(1)*Y6
X13=CL1(1)*Y1-CN1(1)*Y5+CL2(1)*Y2-CN2(1)*Y6
X15=CD1*Y1+CD2*Y2
X17=CF1(1)*Y1-CF2(1)*Y2
X19=CF1(1)*Y5+CF2(1)*Y6
DO 130 J=2,N1
Y=RO(I,J,K)
Y1=U1(I,J)
Y2=U2(I,J)
Y3=V1(I,J)
Y4=V2(I,J)
Y5=UV1(I,J)
Y6=UV2(I,J)
X2=Q1(I,J)*(RB1(J)-W1)+Q2(I,J)*(RB2(J)-W2)
X4=HB1(J)*(P1(I,J)-P1(I-1,J))+HB2(J)*(P2(I,J)-P2(I-1,J))
X6=(CA1(J)*Y3+CA2(J)*Y4)*(RO(I,J+1,K)-Y)
X8=CB1(J)*Y3+CE1(J)*Y1-2.0*CM1(J)*Y5+CB2(J)*Y4+CE2(J)*Y2-2.0*CM2(J
1)*Y6
X10=CC1(J)*Y3-CG1(J)*Y5+CC2(J)*Y4-CG2(J)*Y6
X12=CG1(J)*Y1-CC1(J)*Y5-CG2(J)*Y2+CC2(J)*Y6
X14=CL1(J)*Y1-CN1(J)*Y5+CL2(J)*Y2-CN2(J)*Y6
X16=CD1*Y1+CD2*Y2
X18=CF1(J)*Y1-CF2(J)*Y2
X20=CF1(J)*Y5+CF2(J)*Y6
Z1=(CA1(J)*Y1+CA1(J-1)*U1(I,J-1))*(RO(I,J,K+1)-Y)-(CA2(J)*Y2+CA2(J
1-1)*U2(I,J-1))*(Y-XO(I,J,K-1))
Z2=CA1(J)*Y5*(RO(I,J+1,K+1)-Y)+CA1(J-1)*UV1(I,J-1)*(Y-RO(I,J-1,K+1
1))-CA2(J)*Y6*(XO(I,J+1,K-1)-Y)-CA2(J-1)*UV2(I,J-1)*(Y-XO(I,J-1,K-1
2))
DRO=-A1*(X1+X2)+Y*(A2*(X3+X4)+0.25*(X7+X8)-(X10-X9)/HU2-(X11+X12)/
1HV2)-(X6-X5)/HUU2-Z1/HVV2+(1.0-2.0*Y)*0.25*(X13+X14)-(1.0-Y)*(0.25
2*(X15+X16)+(X17+X18)/HV2-(X20-X19)/HU2)+Z2/HUV2
X1=X2
X3=X4
X5=X6
X7=X8
X9=X10
X11=X12
X13=X14
X15=X16
X17=X18
X19=X20
AA=ABS(DRO)
ERO=AMAX1(ERO,AA)
PAS=DG1*(RO(I,J,K)*DH1-DRO-RA1*XO(I,J,K))
DRO=(PAS-RO(I,J,K))/DT
AA1(1)=AA1(1)+DRO*DRO
```

```
       XO(I,J,K)=RO(I,J,K)
   130 RO(I,J,K)=PAS
       RO(I,1,K)=RO(I,N1,K)
       RO(I,N2,K)=RO(I,2,K)
   140 XO(I,N2,K)=XO(I,2,K)
C      ITERATION FOR MAGNETIC AXIS EQUATION
       W3=0.5*(ZA(K)+ZA(K+1))
       W4=0.5*(ZA(K)+ZN(K-1))
       Y1=RA(K+1)-RA(K)
       Y2=RA(K)-RN(K-1)
       Y3=ZA(K+1)-ZA(K)
       Y4=ZA(K)-ZN(K-1)
       HVV=2.0/(HV*HV)
       DO 150 J=2,N1
       X1=RBV1(J)*XB1(J)+RBU1(J)*XN1(J)-RBV2(J)*XB2(J)-RBU2(J)*XN2(J)
       X2=ZBV1(J)*XB1(J)+ZBU1(J)*XN1(J)-ZBV2(J)*XB2(J)-ZBU2(J)*XN2(J)
       SUMR=SUMR-XP1(J)-XP2(J)+0.5*(XQ1(J)*ZBU1(J)+XQ2(J)*ZBU2(J)-RBU1(J)
      1*XC1(J)-RBU2(J)*XC2(J)-RBV1(J)*XG1(J)-RBV2(J)*XG2(J))-(RB1(J)-W1)*
      2XA1(J)-(RB2(J)-W2)*XA2(J)-HVV*(Y1*XD1(J)-Y2*XD2(J))-(XF1(J)*(0.5*Y
      31+RB1(J)-W1)+XF2(J)*(0.5*Y2-RB2(J)+W2)+X1)/HV
       SUMZ=SUMZ-0.5*(XQ1(J)*RBU1(J)+XQ2(J)*RBU2(J)+ZBU1(J)*XC1(J)+ZBU2(J
      1)*XC2(J)+ZBV1(J)*XG1(J)+ZBV2(J)*XG2(J))-(ZB1(J)-W3)*XA1(J)-(ZB2(J)
      2-W4)*XA2(J)-HVV*(Y3*XD1(J)-Y4*XD2(J))-(XF1(J)*(0.5*Y3+ZB1(J)-W3)+X
      3F2(J)*(0.5*Y4-ZB2(J)+W4)+X2)/HV
   150 CONTINUE
       SUMR=SUMR*HS*HU
       SUMZ=SUMZ*HS*HU
       AA=ABS(SUMR)
       BB=ABS(SUMZ)
       EAX=AMAX1(EAX,AA,BB)
       SUMR=SUMR*FAXIS
       SUMZ=SUMZ*FAXIS
       PAS1=DG3*(RA(K)*DH3-SUMR-RA3*RN(K))
       PAS2=DG3*(ZA(K)*DH3-SUMZ-RA3*ZN(K))
       SUMR=(PAS1-RA(K))/DT
       SUMZ=(PAS2-ZA(K))/DT
       AA1(3)=AA1(3)+SUMR*SUMR+SUMZ*SUMZ
       RN(K)=RA(K)
       ZN(K)=ZA(K)
       RA(K)=PAS1
       ZA(K)=PAS2
   160 CONTINUE
       IF (NPLOT.GT.0) GO TO 200
C      PERIODICITY CONDITIONS
       DO 180 J=1,N2
       DO 170 I=1,NI
       AL(I,J,1)=AL(I,J,N4)-Q(I)
       XL(I,J,1)=AL(I,J,1)
   170 AL(I,J,N5)=AL(I,J,2)+Q(I)
       DO 180 I=2,N3
       RO(I,J,1)=RO(I,J,N4)
       XO(I,J,1)=RO(I,J,1)
   180 RO(I,J,N5)=RO(I,J,2)
       RA(1)=RA(N4)
       RA(N5)=RA(2)
       ZA(1)=ZA(N4)
       X2=HU*HV*ZLE
```

```
C       PRESSURE COMPUTATION
        DO 190 I=1,N3
  190 PR(I)=(AM(I)/(SPR(I)*X2))**GAM
        ZA(N5)=ZA(2)
        RN(1)=RA(1)
        ZN(1)=ZA(1)
C       FUNCTIONALS FOR COMPUTATION OF DESCENT COEFFICIENTS
        AA1(1)=AA1(1)*HS*HU*HV
        AA1(2)=AA1(2)*HS*HU*HV
        AA1(3)=AA1(3)*HV
  200 CONTINUE
        ENER=ENER*HS*HU*HV*ZLE
        RETURN
        END


        SUBROUTINE CBO (K)
C       EVALUATES BOUNDARY QUANTITIES AT K+1/2
        COMMON RO(10,30,30),AL(10,30,30),XO(10,30,30),XL(10,30,30),R(30,30
      1),Z(30,30),RU(30,30),ZU(30,30),RV(30,30),ZV(30,30),X(30,30),RA(30)
      2,ZA(30),RN(30),ZN(30),RB1(30),RB2(30),ZB1(30),ZB2(30),RBU1(30),RBU
      32(30),ZBU1(30),ZBU2(30),RBV1(30),RBV2(30),ZBV1(30),ZBV2(30),HB1(30
      4),HB2(30)
        COMMON /POT/ RVA(30),ZVA(30),BPV(10),BTV(10),PT(10,30,30),PP(10,30
      1,30)
        COMMON /CPL/ NI,NJ,NK,EP,ZLE,GAM,SM,N1,N2,N3,N4,N5,NVAC,HS,HU,HV,P
      1I,RX,RY,E4,A1,A2,A3,A4,A5,A6,HU4,HV4,HUV,IC,IO,SA1,SA2,SA3,SE1,SE2
      2,SE3,DT,RA1,RA2,RA3,RE1,RE2,RE3,ENER,FAXIS,DG1,DG2,DG3,DH1,DH2,DH3
      3,NPLOT,I1
        COMMON /CVA/ NIV,FV1,FV2,NV,NP,OM,PM,HR,H1,H2,H3,C1,C2,N6,EVAC,ETO
      1T,A11,A22,A12,HR4,DTB,FAXV,OM1,OM2,PM1,PM2
        IF (NVAC.LT.0) GO TO 20
C       EVALUATE R AND Z COMPONENTS AND U AND V DERIVATIVES AT THE FREE
C       BOUNDARY
        HU2=2.0*HU
        HV2=2.0*HV
        Y1=(RVA(K)+RVA(K+1))/2.0
        Y2=(ZVA(K)+ZVA(K+1))/2.0
        Y3=(RVA(K+1)-RVA(K))/HV
        Y4=(ZVA(K+1)-ZVA(K))/HV
        DO 10 J=2,N1
        X1=(X(J,K)+X(J,K+1)+X(J+1,K)+X(J+1,K+1))/4.0
        X2=(X(J+1,K)+X(J+1,K+1)-X(J,K)-X(J,K+1))/HU2
        X3=(X(J,K+1)+X(J+1,K+1)-X(J,K)-X(J+1,K))/HV2
        X8=R(J,K)-Y1
        X9=Z(J,K)-Y2
        RB1(J)=Y1+X1*X8
        ZB1(J)=Y2+X1*X9
        RBU1(J)=X2*X8+X1*RU(J,K)
        ZBU1(J)=X2*X9+X1*ZU(J,K)
        RBV1(J)=(1.0-X1)*Y3+X3*X8+X1*RV(J,K)
   10 ZBV1(J)=(1.0-X1)*Y4+X3*X9+X1*ZV(J,K)
        RB1(1)=RB1(N1)
        ZB1(1)=ZB1(N1)
```

```
          RBU1(1)=RBU1(N1)
          ZBU1(1)=ZBU1(N1)
          kBV1(1)=RBV1(N1)
          ZBV1(1)=ZBV1(N1)
          GO TO 40
C         PLASMA BOUNDARY CORRESPONDS TO OUTER WALL
       20 DO 30 J=1,N1
          RB1(J)=R(J,K)
          ZB1(J)=Z(J,K)
          RBU1(J)=RU(J,K)
          ZBU1(J)=ZU(J,K)
          kBV1(J)=RV(J,K)
       30 ZBV1(J)=ZV(J,K)
       40 CONTINUE
          RETURN
          END



          SUBROUTINE CIN (K,BJA,SJA)
C         EVALUATES COEFFICIENTS FOR PLASMA EQUATIONS AT K+1/2
          COMMON RO(10,30,30),AL(10,30,30),XO(10,30,30),XL(10,30,30),R(30,30
         1),Z(30,30),RU(30,30),ZU(30,30),RV(30,30),ZV(30,30),X(30,30),RA(30)
         2,ZA(30),RN(30),ZN(30),RB1(30),RB2(30),ZB1(30),ZB2(30),RBU1(30),RBU
         32(30),ZBU1(30),ZBU2(30),RBV1(30),RBV2(30),ZBV1(30),ZBV2(30),HB1(30
         4),HB2(30)
          COMMON /AUX/ RR2,ZZ2,RRA,ZZA,XX(30),E1(10,30),E2(10,30),F1(10,30),
         1F2(10,30),G1(10,30),G2(10,30),P1(10,30),P2(10,30),Q1(10,30),Q2(10,
         230),V1(10,30),V2(10,30),U1(10,30),U2(10,30),UV1(10,30),UV2(10,30),
         3D(10,30),DD(10,30),BK(10,30),CA1(30),CA2(30),CB1(30),CB2(30),CC1(3
         40),CC2(30),CD1,CD2,CE1(30),CE2(30),CF1(30),CF2(30),CG1(30),CG2(30)
         5,CL1(30),CL2(30),CM1(30),CM2(30),CN1(30),CN2(30),XA1(30),XA2(30),X
         6B1(30),XB2(30),XC1(30),XC2(30),XG1(30),XG2(30),XN1(30),XN2(30),XD1
         7(30),XD2(30),XF1(30),XF2(30),XP1(30),XP2(30),XQ1(30),XQ2(30),SPR(1
         80)
          COMMON /INP/ Q(10),QT(10),AM(10),PR(10),QQ(10),PC(10),TC(10),BPP(1
         10),BTP(10),BET(10)
          COMMON /AC/ EN(100,5),ET(5),AA1(5),AA2(5),AA3(5),AVE(5),NAC,NE,SAF
         1I,SAFV,SAFPSI,SAFRO,SAFAX
          COMMON /CPL/ NI,NJ,NK,EP,ZLE,GAM,SM,N1,N2,N3,N4,N5,NVAC,HS,HU,HV,P
         1I,RX,RY,E4,A1,A2,A3,A4,A5,A6,HU4,HV4,HUV,IC,IO,SA1,SA2,SA3,SE1,SE2
         2,SE3,DT,RA1,RA2,RA3,RE1,RE2,RE3,ENER,FAXIS,DG1,DG2,DG3,DH1,DH2,DH3
         3,NPLOT,I1
          COMMON /CVA/ NIV,FV1,FV2,NV,NP,OM,PM,HR,H1,H2,H3,C1,C2,N6,EVAC,ETO
         1T,A11,A22,A12,HR4,DTB,FAXV,OM1,OM2,PM1,PM2
          ZZ=ZLE*ZLE
          ZL2=1.0/(2.0*ZZ)
          ZL4=1.0/(4.0*ZZ)
          GAM1=GAM-1.0
          HVV=2.0*HV*HV
          HUU=2.0*HU*HU
          HU2=2.0*HU
          HV2=2.0*HV
          Y1=(RA(K)+RA(K+1))/2.0
          Y2=(ZA(K)+ZA(K+1))/2.0
```

```
        Y3=(RA(K+1)-RA(K))/HV
        Y4=(ZA(K+1)-ZA(K))/HV
C       COMPUTE COEFFICIENTS FOR R COMING FROM BOUNDARY QUANTITIES
        CD1=Y3*Y3+Y4*Y4
        DO 10 J=1,N1
        X1=RB1(J)-Y1
        X2=ZB1(J)-Y2
        HB1(J)=X1*ZBU1(J)-X2*RBU1(J)
        CA1(J)=X1*X1+X2*X2
        CB1(J)=RBU1(J)*RBU1(J)+ZBU1(J)*ZBU1(J)
        CC1(J)=X1*RBU1(J)+X2*ZBU1(J)
        CE1(J)=RBV1(J)*RBV1(J)+ZBV1(J)*ZBV1(J)
        CF1(J)=X1*Y3+X2*Y4
        CG1(J)=X1*RBV1(J)+X2*ZBV1(J)
        CL1(J)=RBV1(J)*Y3+ZBV1(J)*Y4
        CM1(J)=RBU1(J)*RBV1(J)+ZBU1(J)*ZBV1(J)
     10 CN1(J)=Y3*RBU1(J)+Y4*ZBU1(J)
C       COMPUTE JACOBIAN
        DO 20 I=1,N3
        X1=RO(I+1,2,K)*RO(I+1,2,K)-RO(I,2,K)*RO(I,2,K)+RO(I+1,2,K+1)*RO(I+
     11,2,K+1)-RO(I,2,K+1)*RO(I,2,K+1)
        DO 20 J=2,N1
        X2=RO(I+1,J+1,K)*RO(I+1,J+1,K)-RO(I,J+1,K)*RO(I,J+1,K)+RO(I+1,J+1,
     1K+1)*RO(I+1,J+1,K+1)-RO(I,J+1,K+1)*RO(I,J+1,K+1)
        D(I,J)=HB1(J)*A6*(X1+X2)
        BJA=AMIN1(BJA,D(I,J))
        SJA=AMAX1(SJA,D(I,J))
     20 X1=X2
        IF (BJA.GT.0.0) GO TO 30
        PRINT 110
        STOP
C       PSI DERIVATIVES AND TOROIDAL FACTOR
     30 DO 40 I=1,NI
        X1=(AL(I,2,K+1)-AL(I,2,K))*(AL(I,2,K+1)-AL(I,2,K))
        DO 40 J=2,N1
        X2=(AL(I,J+1,K+1)-AL(I,J+1,K))*(AL(I,J+1,K+1)-AL(I,J+1,K))
        V1(I,J)=(X1+X2)/HVV
        X1=X2
        U1(I,J)=((AL(I,J+1,K+1)-AL(I,J,K+1))*(AL(I,J+1,K+1)-AL(I,J,K+1))+(
     1AL(I,J+1,K)-AL(I,J,K))*(AL(I,J+1,K)-AL(I,J,K)))/HUU
        UV1(I,J)=((AL(I,J+1,K+1)-AL(I,J,K))*(AL(I,J+1,K+1)-AL(I,J,K))-(AL(
     1I,J+1,K)-AL(I,J,K+1))*(AL(I,J+1,K)-AL(I,J,K+1)))*HUV
     40 BK(I,J)=1.0+EP*(Y1+0.25*(RB1(J)-Y1)*(RO(I,J,K)+RO(I,J+1,K)+RO(I,J,
     1K+1)+RO(I,J+1,K+1)))
        DO 50 J=2,N1
        DD(1,J)=1.0/D(1,J)
        DD(NI,J)=1.0/D(N3,J)
        DO 50 I=2,N3
     50 DD(I,J)=1.0/D(I,J)+1.0/D(I-1,J)
        DO 60 J=2,N1
        XA1(J)=0.0
        XB1(J)=0.0
        XC1(J)=0.0
        XD1(J)=0.0
        XG1(J)=0.0
        XN1(J)=0.0
        XF1(J)=0.0
```

```
      DO 60 I=1,NI
      Y1=RO(I,J,K)
      Y2=RO(I,J+1,K)
      Y3=RO(I,J,K+1)
      Y4=RO(I,J+1,K+1)
      Y5=1.0-Y1
      Y6=1.0-Y2
      Y7=1.0-Y3
      Y8=1.0-Y4
C     MAPPING DERIVATIVES
      X1=((Y2-Y1)*(Y2-Y1)+(Y4-Y3)*(Y4-Y3))/HUU
      X2=0.25*(Y1*Y1+Y2*Y2+Y3*Y3+Y4*Y4)
      X3=(Y2*Y2-Y1*Y1+Y4*Y4-Y3*Y3)/HU2
      RU2=CA1(J)*X1+CB1(J)*X2+CC1(J)*X3
      X4=0.25*(Y5*Y5+Y6*Y6+Y7*Y7+Y8*Y8)
      X5=((Y3-Y1)*(Y3-Y1)+(Y4-Y2)*(Y4-Y2))/HVV
      X6=(Y7*Y7-Y5*Y5+Y8*Y8-Y6*Y6)/HV2
      X7=0.25*(Y1*Y5+Y2*Y6+Y3*Y7+Y4*Y8)
      X8=(Y3*Y3-Y1*Y1+Y4*Y4-Y2*Y2)/HV2
      RV2=CD1*X4+CA1(J)*X5+CE1(J)*X2-CF1(J)*X6+CG1(J)*X8+2.0*X7*CL1(J)+Z
     1Z*BK(I,J)*BK(I,J)
      X9=(Y6*Y6-Y5*Y5+Y8*Y8-Y7*Y7)/HU2
      X10=HUV*((Y4-Y1)*(Y4-Y1)-(Y2-Y3)*(Y2-Y3))
      RUV=0.5*(-CF1(J)*X9+CC1(J)*X8+CG1(J)*X3)+CH1(J)*X2+CA1(J)*X10+CN1(
     1J)*X7
      ALF=V1(I,J)*RU2+U1(I,J)*RV2-2.0*UV1(I,J)*RUV
      ALF1=(ALF*ZL2)/(BK(I,J)*BK(I,J))
      Q1(I,J)=(ALF1-U1(I,J))*DD(I,J)
      P1(I,J)=(ALF*ZL4)/BK(I,J)
      FAC=DD(I,J)/BK(I,J)
      FAC1=FAC*ZL2
C     COMPUTE COEFFICIENTS FOR R COMING FROM PSI DERIVATIVES
      U1(I,J)=U1(I,J)*FAC1
      V1(I,J)=V1(I,J)*FAC1
      UV1(I,J)=UV1(I,J)*FAC1
C     COMPUTE COEFFICIENTS FOR PSI EQUATION
      E1(I,J)=RU2*FAC
      F1(I,J)=RUV*FAC
      G1(I,J)=RV2*FAC
C     COMPUTE COEFFICIENTS FOR AXIS EQUATIONS
      XA1(J)=XA1(J)+0.5*(V1(I,J)*X1+U1(I,J)*X5-2.0*UV1(I,J)*X10)
      XC1(J)=XC1(J)+0.5*(V1(I,J)*X3-UV1(I,J)*X8)
      XG1(J)=XG1(J)+0.5*(U1(I,J)*X8-UV1(I,J)*X3)
      XB1(J)=XB1(J)+U1(I,J)*X7
      XN1(J)=XN1(J)-UV1(I,J)*X7
      XD1(J)=XD1(J)+0.5*U1(I,J)*X4
   60 XF1(J)=XF1(J)+0.5*(UV1(I,J)*X9-U1(I,J)*X6)
C     COMPUTE ENERGY AND REMAINING COEFFICIENTS FOR R AND AXIS
      DO 80 J=2,N1
      XQ1(J)=0.0
      DO 70 I=1,N3
      X3=0.5*(BK(I,J)+BK(I+1,J))
      X1=PR(I)*X3
      X2=(P1(I,J)+P1(I+1,J))/D(I,J)
      ENER=ENER+X2+X1*D(I,J)/GAM1
      P1(I,J)=X2/D(I,J)+X1
      XQ1(J)=XQ1(J)+P1(I,J)*D(I,J)
```

```
   70 SPR(I)=SPR(I)+X3*D(I,J)
      XQ1(J)=XQ1(J)/HB1(J)
      Q1(1,J)=Q1(1,J)+PR(1)*D(1,J)
      Q1(NI,J)=Q1(NI,J)+PR(N3)*D(N3,J)
      DO 80 I=2,N3
   80 Q1(I,J)=Q1(I,J)+PR(I)*D(I,J)+PR(I-1)*D(I-1,J)
      DO 90 J=2,N1
      XP1(J)=0.0
      E1(NI,J)=2.0*E1(NI,J)
      F1(NI,J)=2.0*F1(NI,J)
      G1(NI,J)=2.0*G1(NI,J)
      DO 90 I=1,NI
   90 XP1(J)=XP1(J)+E4*Q1(I,J)*(1.0-0.25*(RO(I,J,K)+RO(I,J+1,K)+RO(I,J,K
     1+1)+RO(I,J+1,K+1)))
C     PERIODICITY CONDITIONS
      XA1(1)=XA1(N1)
      XB1(1)=XB1(N1)
      XC1(1)=XC1(N1)
      XG1(1)=XG1(N1)
      XD1(1)=XD1(N1)
      XF1(1)=XF1(N1)
      XN1(1)=XN1(N1)
      XP1(1)=XP1(N1)
      XQ1(1)=XQ1(N1)
      DO 100 I=1,NI
      U1(I,1)=U1(I,N1)
      V1(I,1)=V1(I,N1)
      UV1(I,1)=UV1(I,N1)
      E1(I,1)=E1(I,N1)
      F1(I,1)=F1(I,N1)
      G1(I,1)=G1(I,N1)
      P1(I,1)=P1(I,N1)
  100 Q1(I,1)=Q1(I,N1)
      RETURN
  110 FORMAT (////,6X17HNEGATIVE JACOBIAN)
      END



      SUBROUTINE TPLOT (ITER)
C     FINAL PRINT OUT AND PLOTS. THIS USES SUBROUTINES PLOT, AXIS,
C     NUMBER AND SYMBOL, WHICH ARE PART OF THE SCOOP SOFTWARE PACKAGE
C     FOR THE CALCOMP PLOTTER AT NYU
      COMMON /FUNC/ ALF,RBOU,DELO,DEL1,DEL2,DEL3,DEL10,DEL20,DEL30,DEL22
     1,DEL33,PO,XPR,AMUO,AMU1,AMU2,AMP,FUR(7),FUZ(7),FUX(7),NRUN
      COMMON RO(10,30,30),AL(10,30,30),XO(10,30,30),XL(10,30,30),R(30,30
     1),Z(30,30),RU(30,30),ZU(30,30),RV(30,30),ZV(30,30),X(30,30),RA(30)
     2,ZA(30),RN(30),ZN(30),RB1(30),RB2(30),ZB1(30),ZB2(30),RBU1(30),RBU
     32(30),ZBU1(30),ZBU2(30),RBV1(30),RBV2(30),ZBV1(30),ZBV2(30),HB1(30
     4),HB2(30)
      COMMON /AUX/ SP(100),SD(100),SI(100),SJ(100),SIV(50),SJV(50),SPP(5
     10),SVV(50),RP(50),RM(50),RP1(50),RV1(50),F(100),FP(100),FPP(100),F
     2PPP(100),XOR(4),YOR(4),YB(4,4),SLRV1(50),SIRV1(50),SARV1(50),SLRV2
     3(50),SIRV2(50),SARV2(50),SLZV1(50),SIZV1(50),SAZV1(50),SLZV2(50),S
     4IZV2(50),SAZV2(50),SLRA1(50),SIRA1(50),SARA1(50),SLRA2(50),SIRA2(5
```

```
      50),SARA2(50),SLZA1(50),SIZA1(50),SAZA1(50),SLZA2(50),SIZA2(50),SAZ
      6A2(50),SLMK1(50),SIMK1(50),SAMK1(50),SLMK2(50),SIMK2(50),SAMK2(50)
      7,SLMK3(50),SIMK3(50),SAMK3(50),SLMK4(50),SIMK4(50),SAMK4(50),NITER
      8(50)
      COMMON /POT/ RVA(30),ZVA(30),BPV(10),BTV(10),PT(10,30,30),PP(10,30
     1,30)
      COMMON /INP/ Q(10),QT(10),AM(10),PR(10),QQ(10),PC(10),TC(10),BPP(1
     10),BTP(10),BET(10)
      COMMON /AC/ EN(100,5),ET(5),AA1(5),AA2(5),AA3(5),AVE(5),NAC,NE,SAF
     1I,SAFV,SAFPSI,SAFRO,SAFAX
      COMMON /CPL/ NI,NJ,NK,EP,ZLE,GAM,SM,N1,N2,N3,N4,N5,NVAC,HS,HU,HV,P
     1I,RX,RY,E4,A1,A2,A3,A4,A5,A6,HU4,HV4,HUV,IC,IO,SA1,SA2,SA3,SE1,SE2
     2,SE3,DT,RA1,RA2,RA3,RE1,RE2,RE3,ENER,FAXIS,DG1,DG2,DG3,DH1,DH2,DH3
     3,NPLOT,I1
      COMMON /CVA/ NIV,FV1,FV2,NV,NP,OM,PM,HR,H1,H2,H3,C1,C2,N6,EVAC,ETO
     1T,A11,A22,A12,HR4,DTB,FAXV,OM1,OM2,PM1,PM2
      COMMON /FOU/ SV(7,30),SU(7,30),SFI(7,30),SRO(7,7),XR(7),XZ(7),YR(7
     1),YZ(7)
      COMMON /PLOT/ NRA1,NRA2,NZA1,NZA2,MK1,MK2,MK3,MK4,M1,M2,M3,M4,K1,K
     12,K3,K4,RNAME(7),ZNAME(7)
      DIMENSION PL1(1500), PL2(1500), PL3(1500), PL4(1500), PL5(1500)
      DIMENSION XPL(1500,5)
      EQUIVALENCE (RO(1,1,1),XPL(1,1))
      EQUIVALENCE (XPL(1,1),PL1(1)), (XPL(1,2),PL2(1)), (XPL(1,3),PL3(1)
     1), (XPL(1,4),PL4(1)), (XPL(1,5),PL5(1))
      DIMENSION TEXTX(3), TEXTY(3), SYM(4), T(7), THE(4)
C     INITIALIZE PLOT PACKAGE
      DO 10 I=1,7
   10 T(I)=0.
      ENCODE (27,460,T) NRUN
      CALL PLOTSBL (300,T)
      CALL PLOT (1.0,1.0,-3)
      GO TO 140
   20 CONTINUE
C     INTERPOLATION AND PLOT OF PRESSURE AND MU
      DR=RP(NI)/49.0
      DO 30 I=1,50
   30 SP(I)=(I-1)*DR
      IND=10
      CALL SPLIF (1,NI,RP1,QQ,FP,FPP,FPPP,3,0.,3,0.,0,0.0,IND)
      CALL INTPL (1,50,SP,SI,1,NI,RP1,QQ,FP,FPP,FPPP,0)
      CALL SPLIF(1,NI,RP1,PR,FP,FPP,FPPP,2,0.,3,0.,0,0.,IND)
      CALL INTPL (1,50,SP,SJ,1,NI,RP1,PR,FP,FPP,FPPP,0)
      CALL PLOT (1.0,7.0,-3)
      TEXTX(1)=10H
      TEXTX(2)=10H        RADIU
      TEXTX(3)=10HS
      TEXTY(1)=10H
      TEXTY(2)=10H        PRESSU
      TEXTY(3)=10HRE
      SYM(1)=10H
      SYM(2)=10H
      SYM(3)=10H
      SYM(4)=10H
      DO 40 I=1,50
      PL1(I)=SP(I)
   40 PL2(I)=SJ(I)
```

```
          NPLO=50
          IF (NVAC.LT.0) GO TO 60
          DR=(RM(NIV)-RM(1))/49.0
          DO 50 I=1,50
          SD(I)=RM(1)+(I-1)*DR
          PL1(I+50)=SD(I)
       50 PL2(I+50)=0.
          NPLO=100
       60 CALL PLOTB (2.,2.,NPLO,1,TEXTX,TEXTY,SYM)
          CALL PLOT (3.5,0.,-3)
          TEXTY(2)=10H        MU
          TEXTY(3)=10H
          DO 70 I=1,50
       70 PL2(I)=SI(I)
          CALL PLOTB (2.0,2.0,50,1,TEXTX,TEXTY,SYM)
C         INTERPOLATION AND PLOT OF MAGNETIC FIELD
          CALL SPLIF (1,NI,RP1,BTP,FP,FPP,FPPP,3,0.,3,0.,0,0.0,IND)
          CALL INTPL (1,50,SP,SI,1,NI,RP1,BTP,FP,FPP,FPPP,0)
          CALL SPLIF (1,NI,RP1,BPP,FP,FPP,FPPP,3,0.,3,0.,0,0.0,IND)
          CALL INTPL (1,50,SP,SJ,1,NI,RP1,BPP,FP,FPP,FPPP,0)
          CALL PLOT (-3.5,-4.0,-3)
          TEXTY(2)=10H MAGNETIC
          TEXTY(3)=10H FIELD
          SYM(1)=10HBT
          SYM(2)=10HBP
          DO 80 I=1,50
          PL2(I)=SI(I)
       80 PL3(I)=SJ(I)
          NPLO=50
          IF (NVAC.LT.0) GO TO 100
          CALL SPLIF (1,N6,RV1,BTV,FP,FPP,FPPP,3,0.,3,0.,0,0.,IND)
          CALL INTPL (1,50,SD,SIV,1,N6,RV1,BTV,FP,FPP,FPPP,0)
          CALL SPLIF (1,N6,RV1,BPV,FP,FPP,FPPP,3,0.,3,0.,0,0.,IND)
          CALL INTPL (1,50,SD,SJV,1,N6,RV1,BPV,FP,FPP,FPPP,0)
          DO 90 I=1,50
          PL2(I+50)=SIV(I)
       90 PL3(I+50)=SJV(I)
          NPLO=100
      100 CALL PLOTB (2.0,2.0,NPLO,2,TEXTX,TEXTY,SYM)
C         INTERPOLATION AND PLOT OF CURRENT
          CALL PLOT (3.5,0.,-3)
          TEXTY(2)=10H        CURREN
          TEXTY(3)=10HT
          SYM(1)=10HIT
          SYM(2)=10HIP
          CALL SPLIF (1,NI,RP1,TC,FP,FPP,FPPP,3,0.,3,0.,0,0.0,IND)
          CALL INTPL (1,50,SP,SI,1,NI,RP1,TC,FP,FPP,FPPP,0)
          CALL SPLIF (1,NI,RP1,PC,FP,FPP,FPPP,3,0.,3,0.,0,0.0,IND)
          CALL INTPL (1,50,SP,SJ,1,NI,RP1,PC,FP,FPP,FPPP,0)
          DO 110 I=1,50
          PL2(I)=SI(I)
      110 PL3(I)=SJ(I)
          IF (NVAC.LT.0) GO TO 130
          DO 120 I=1,50
          PL2(I+50)=C1
      120 PL3(I+50)=C2
      130 CALL PLOTB (2.0,2.0,NPLO,2,TEXTX,TEXTY,SYM)
```

```
      GO TO 320
C     CROSS SECTIONS PLOT
  140 XOR(1)=2.0
      XOR(2)=5.7
      XOR(3)=2.0
      XOR(4)=5.7
      YOR(1)=7.5
      YOR(2)=7.5
      YOR(3)=3.5
      YOR(4)=3.5
      DF=1.0/99.0
      DO 150 J=2,N2
  150 SP(J)=(J-2)*HU
      DO 160 I=1,100
  160 SD(I)=(I-1)*DF
      XY=FLOAT(NK)/4.0
      NN=INT(XY)
      AA=0.0
      DO 170 L=1,4
      K=2+INT(FLOAT(L-1)*XY)
      DO 170 J=2,N1
      CC=ABS(R(J,K))
      DD=ABS(Z(J,K))
  170 AA=AMAX1(AA,CC,DD)
      SCA=1.4/AA
      IPL=0
      DO 280 L=1,4
      K=2+INT(FLOAT(L-1)*XY)
      IPL=IPL+1
      IF (IPL.GT.4) GO TO 290
      THE(IPL)=FLOAT(K-2)/FLOAT(NK)+HV/2.
      CALL PLOT (XOR(IPL),YOR(IPL),-3)
      CALL PLOT (-1.5,0.,3)
      CALL PLOT (1.5,0.,2)
      CALL PLOT (0.,-1.5,3)
      CALL PLOT (0.,1.5,2)
      DO 220 I=2,NI
      DO 180 J=2,N1
      X1=0.5*(RVA(K)+RVA(K+1))
      X2=0.25*(X(J,K)+X(J,K+1)+X(J+1,K)+X(J+1,K+1))
      X3=0.25*(RO(I,J,K)+RO(I,J+1,K)+RO(I,J,K+1)+RO(I,J+1,K+1))
      X4=0.5*(RA(K)+RA(K+1))
      X1=X1+X2*(R(J,K)-X1)
  180 F(J)=X4+X3*(X1-X4)
      F(1)=F(N1)
      F(N2)=F(2)
      VM=(F(3)+F(1)-2.0*F(2))/(HU*HU)
      CALL SPLIF (2,N2,SP,F,FP,FPP,FPPP,2,VM,2,VM,0,0.0,IND)
      CALL INTPL (1,100,SD,SI,2,N2,SP,F,FP,FPP,FPPP,0)
      DO 190 J=2,N1
      X1=0.5*(ZVA(K)+ZVA(K+1))
      X2=0.25*(X(J,K)+X(J,K+1)+X(J+1,K)+X(J+1,K+1))
      X3=0.25*(RO(I,J,K)+RO(I,J+1,K)+RO(I,J,K+1)+RO(I,J+1,K+1))
      X4=0.5*(ZA(K)+ZA(K+1))
      X1=X1+X2*(Z(J,K)-X1)
  190 F(J)=X4+X3*(X1-X4)
      F(1)=F(N1)
```

```
      F(N2)=F(2)
      VM=(F(3)+F(1)-2.0*F(2))/(HU*HU)
      CALL SPLIF (2,N2,SP,F,FP,FPP,FPPP,2,VM,2,VM,0,0.0,IND)
      CALL INTPL (1,100,SD,SJ,2,N2,SP,F,FP,FPP,FPPP,0)
      DO 200 J=1,100
      SI(J)=SI(J)*SCA
  200 SJ(J)=SJ(J)*SCA
      CALL PLOT (SI(1),SJ(1),3)
      DO 210 J=1,100
  210 CALL PLOT (SI(J),SJ(J),2)
  220 CONTINUE
      IF (NVAC.LT.0) GO TO 270
      DO 230 J=2,N1
  230 F(J)=R(J,K)
      F(1)=F(N1)
      F(N2)=F(2)
      VM=(F(3)+F(1)-2.0*F(2))/(HU*HU)
      CALL SPLIF (2,N2,SP,F,FP,FPP,FPPP,2,VM,2,VM,0,0.0,IND)
      CALL INTPL (1,100,SD,SI,2,N2,SP,F,FP,FPP,FPPP,0)
      DO 240 J=2,N1
  240 F(J)=Z(J,K)
      F(1)=F(N1)
      F(N2)=F(2)
      VM=(F(3)+F(1)-2.0*F(2))/(HU*HU)
      CALL SPLIF (2,N2,SP,F,FP,FPP,FPPP,2,VM,2,VM,0,0.0,IND)
      CALL INTPL (1,100,SD,SJ,2,N2,SP,F,FP,FPP,FPPP,0)
      DO 250 J=1,100
      SI(J)=SI(J)*SCA
  250 SJ(J)=SJ(J)*SCA
      CALL PLOT (SI(1),SJ(1),3)
      DO 260 J=1,100
  260 CALL PLOT (SI(J),SJ(J),2)
  270 CONTINUE
      XRA=0.5*(RA(K)+RA(K+1))*SCA-0.02857
      ZRA=0.5*(ZA(K)+ZA(K+1))*SCA-0.05
      CALL SYMBOL (XRA,ZRA,0.1,1HX,0.0,1)
      XRA=-XOR(IPL)
      ZRA=-YOR(IPL)
  280 CALL PLOT (XRA,ZRA,-3)
  290 CONTINUE
      RINR=RX
      ASP=ZLE/(2.0*PI*RX)
      IF (EP.LT.0.00001) GO TO 300
      ENCODE (54,470,T) THE(1),THE(2),THE(3),THE(4),ASP
      CALL SYMBOL (0.5,1.25,0.14,T,0.,54)
      RAYR=1./EP
      ENCODE (54,480,T) RAYR,RINR
      GO TO 310
  300 ENCODE (54,490,T) THE(1),THE(2),THE(3),THE(4),ASP
      CALL SYMBOL (0.5,1.25,0.14,T,0.,54)
      ENCODE (54,500,T) RINR
  310 CALL SYMBOL (0.5,.9,0.14,T,0.,54)
      CALL PLOT (12.0,0.0,-3)
      GO TO 20
C     ENERGY PLOT
  320 REWIND 3
      REWIND 4
```

```
      XY=FLOAT(ITER)/FLOAT(IC)
      N=INT(XY-1.0)
      IF (N.GT.1500) N=1500
      CALL PLOT (9.5,3.0,-3)
      DO 330 I=1,N
      PL1(I)=(I-1)*DT*IC
      READ (3) PL2(I),ERO,EAL,EAX
      READ (3) (XR(M),M=1,7),(KZ(M),M=1,7)
      READ (3) ((YB(J,K),K=1,4),J=1,4)
  330 READ (3) EA1,EA2,EA3
      REWIND 3
      TEXTX(1)=10H          A
      TEXTX(2)=10HRTIFICIAL
      TEXTX(3)=10HTIME
      TEXTY(1)=10H
      TEXTY(2)=10HLOG10 ENER
      TEXTY(3)=10HGY
      SYM(1)=10H
      PLL=PL2(1)
      DO 340 I=1,N
  340 PL2(I)=ALOG10(PL2(I)/PLL)
      CALL PLOTB (3.0,3.0,N,1,TEXTX,TEXTY,SYM)
      CALL PLOT (0.,-4.0,-3)
C     PLASMA RESIDUALS PLOT
      DO 350 I=1,N
      READ (3) ETOT,ERO,EAL,EAX
      READ (3) (XR(M),M=1,7),(XZ(M),M=1,7)
      READ (3) ((YB(J,K),K=1,4),J=1,4)
      READ (3) EA1,EA2,EA3
      PL2(I)=ALOG10(ERO)
      PL3(I)=ALOG10(EAL)
      PL4(I)=ALOG10(EAX)
  350 CONTINUE
      REWIND 3
      TEXTY(1)=10H          LOG
      TEXTY(2)=10H10 RESIDUA
      TEXTY(3)=10HL PLASMA
      SYM(1)=10HR
      SYM(2)=10HPSI
      SYM(3)=10HAXIS
      CALL PLOTB (3.0,3.0,N,3,TEXTX,TEXTY,SYM)
      CALL PLOT (12.,4.0,-3)
      IF (NVAC.LT.0) GO TO 380
C     VACUUM RESIDUALS PLOT
      DO 360 I=1,N
      READ (4) ERBO,ERBO1,ERVA
      READ (4) (YR(M),M=1,7),(YZ(M),M=1,7)
      READ (4) OM1,OM2
      PL2(I)=ALOG10(ERBO)
      PL3(I)=ALOG10(ERBO1)
      PL4(I)=ALOG10(ERVA)
  360 CONTINUE
      REWIND 4
      TEXTY(3)=10HL VACUUM
      SYM(1)=10HBOUNDARY
      SYM(2)=10HAXIS
      SYM(3)=10HVACUUM
```

```
      CALL PLOTB (3.,3.,N,3,TEXTX,TEXTY,SYM)
      CALL PLOT (0.,-4.0,-3)
C     PLOT OF FOURIER COEFFICIENTS FOR THE VACUUM AXIS
      DO 370 I=1,N
      READ (4) ERBO,ERBO1,ERVA
      READ (4) (YR(M),M=1,7),(YZ(M),M=1,7)
      READ (4) OM1,OM2
      PL2(I)=YR(NRA1)
      PL3(I)=YR(NRA2)
      PL4(I)=YZ(NZA1)
      PL5(I)=YZ(NZA2)
  370 CONTINUE
      REWIND 4
      TEXTY(1)=10HFOURIER CO
      TEXTY(2)=10HEFFICIENT
      TEXTY(3)=10HVAC AXIS
      SYM(1)=RNAME(NRA1)
      SYM(2)=RNAME(NRA2)
      SYM(3)=ZNAME(NZA1)·
      SYM(4)=ZNAME(NZA2)
      CALL PLOTB (3.,3.,N,4,TEXTX,TEXTY,SYM)
      REWIND 4
      CALL PLOT (12.0,4.0,-3)
  360 CONTINUE
C     PLOT OF FOURIER COEFFICIENTS FOR THE MAGNETIC AXIS
      DO 390 I=1,N
      READ (3) ETOT,ERO,EAL,EAX
      READ (3) (XR(M),M=1,7),(XZ(M),M=1,7)
      READ (3) ((YB(J,K),K=1,4),J=1,4)
      READ (3) EA1,EA2,EA3
      PL2(I)=XR(NRA1)
      PL3(I)=XR(NRA2)
      PL4(I)=XZ(NZA1)
      PL5(I)=XZ(NZA2)
  390 CONTINUE
      REWIND 3
      TEXTY(1)=10H  FOURIER
      TEXTY(2)=10HCOEFFICIEN
      TEXTY(3)=10HT MAG AXIS
      SYM(1)=RNAME(NRA1)
      SYM(2)=RNAME(NRA2)
      SYM(3)=ZNAME(NZA1)
      SYM(4)=ZNAME(NZA2)
      CALL PLOTB (3.,3.,N,4,TEXTX,TEXTY,SYM)
      CALL PLOT (0.,-4.5,-3)
C     PLOT OF FOURIER COEFFICIENTS FOR THE FLUX SURFACE
      DO 400 I=1,N
      READ (3) ETOT,ERO,EAL,EAX
      READ (3) (XR(M),M=1,7),(XZ(M),M=1,7)
      READ (3) ((YB(J,K),J=1,4),K=1,4)
      READ (3) EA1,EA2,EA3
      PL2(I)=YB(M1+1,K1+1)
      PL3(I)=YB(M2+1,K2+1)
      PL4(I)=YB(M3+1,K3+1)
      PL5(I)=YB(M4+1,K4+1)
  400 CONTINUE
      REWIND 3
```

```
      IF (NVAC.LT.0) GO TO 410
      TEXTY(1)=10HFOURIER CO
      TEXTY(2)=10HEFFICIENT
      TEXTY(3)=10HFREE BOUND
      GO TO 420
  410 TEXTY(1)=10H  FOURIER
      TEXTY(2)=10HCOEFFICIEN
      TEXTY(3)=10HT RADIUS
  420 TSYM=3HMK=
      ENCODE (10,510,SYM(1) )TSYM,M1,K1
      ENCODE (10,510,SYM(2) )TSYM,M2,K2
      ENCODE (10,510,SYM(3) )TSYM,M3,K3
      ENCODE (10,510,SYM(4) )TSYM,M4,K4
      CALL PLOTB (3.,3.,N,4,TEXTX,TEXTY,SYM)
      CALL PLOT (12.,4.0,-3)
C     PLOT OF DESCENT COEFFICIENTS
      DO 430 I=1,N
      READ (3) ETOT,ERO,EAL,EAX
      READ (3) (XR(M),M=1,7),(XZ(M),M=1,7)
      READ (3) ((YB(J,K),K=1,4),J=1,4)
      READ (3) PL2(I),PL3(I),PL4(I)
  430 CONTINUE
      TEXTY(1)=10H        DE
      TEXTY(2)=10HSCENT COEF
      TEXTY(3)=10HFICIENTS
      SYM(1)=10HE1/A1
      SYM(2)=10HE2/A2
      SYM(3)=10HE3/A3
      CALL PLOTB (3.,3.,N,3,TEXTX,TEXTY,SYM)
      IF (NVAC.LT.0) GO TO 450
C     PLOT OF RELAXATION FACTORS
      DO 440 I=1,N
      READ (4) ERBO,ERBO1,ERVA
      READ (4) (YR(M),M=1,7),(YZ(M),M=1,7)
      READ (4) PL2(I),PL3(I)
  440 CONTINUE
      TEXTY(1)=10H
      TEXTY(2)=10HRELAXATION
      TEXTY(3)=10H FACTORS
      SYM(1)=10HOM1
      SYM(2)=10HOM2
      CALL PLOT (0.,-4.0,-3)
      CALL PLOTB (3.,3.,N,2,TEXTX,TEXTY,SYM)
  450 CALL PLOT (0.0,0.0,999)
      RETURN
  460 FORMAT (25HI.D. NUMBER 110801   RUN=I2)
  470 FORMAT (21HCROSS SECTIONS AT V= F3.2,1H,F3.2,1H,F3.2,1H,F3.2,13H,
     11/(EP*QLZ)=F5.2)
  480 FORMAT (13HMAJOR RADIUS=F6.2,17X13HMINOR RADIUS=F5.2)
  490 FORMAT (21HCROSS SECTIONS AT V= F3.2,1H,F3.2,1H,F3.2,1H,F3.2,11H,
     1QLZ/2*PI=F5.2)
  500 FORMAT (21HMAJOR RADIUS INFINITE,12X14H MINOR RADIUS=F5.2)
  510 FORMAT (A3,2I1)
      END
```

```
      SUBROUTINE PLOTB (XL,YL,N,M,TEXTX,TEXTY,SYM)
C     PLOTS XPL(I,J) FOR 1.LT.J AND J.LE.5 AS A FUNCTION OF XPL(I,1).
C     XL AND YL ARE THE LENGTHS IN INCHES OF THE X AND Y AXES. N IS THE
C     NUMBER OF POINTS PLOTTED FOR EACH CURVE, M IS THE NUMBER OF
C     CURVES. TEXTX AND TEXTY ARE THE LABELS FOR THE X AND Y AXES.
C     SYM(J) IS THE LABEL FOR THE J-TH CURVE.
      COMMON XPL(1500,5)
      DIMENSION TEXTX(3), TEXTY(3), XA(2), XB(2), SYM(4)
      XMIN=1000.
      XMAX=-1000.
      MM=M+1
      YMAX=-1000.
      YMIN=1000.
      DO 10 I=1,N
      XMAX=AMAX1(XMAX,XPL(I,1))
      XMIN=AMIN1(XMIN,XPL(I,1))
      DO 10 J=2,MM
      YMAX=AMAX1(YMAX,XPL(I,J))
   10 YMIN=AMIN1(YMIN,XPL(I,J))
      IF (YMAX.EQ.0.) GO TO 20
      IF ((YMAX-YMIN)/ABS(YMAX).LT.1.E-5) YMIN=YMAX
   20 CONTINUE
      IF (ABS(YMIN).LT.0.01*ABS(YMAX)) YMIN=0.01*ABS(YMAX)
      XA(1)=XMIN
      XA(2)=XMAX
      CALL SCALE (XA,XL,2,1,XO,DX)
      XB(1)=YMIN
      XB(2)=YMAX
      CALL SCALE (XB,YL,2,1,YO,DY)
      CALL AXIS (0.0,0.0,TEXTX,-30,XL,0.,XO,DX)
      CALL AXIS (0.,0.,TEXTY,30,YL,90.,YO,DY)
      DO 30 I=1,N
      DO 30 J=2,MM
   30 XPL(I,J)=(XPL(I,J)-YO)/DY
      DO 50 J=2,MM
      XPL1=(XPL(1,1)-XO)/DX
      CALL PLOT (XPL1,XPL(1,J),3)
      DO 40 I=1,N
      XPL1=(XPL(I,1)-XO)/DX
   40 CALL PLOT (XPL1,XPL(I,J),2)
      CALL SYMBOL (XPL1+.1,XPL(N,J),0.1,SYM(J-1),0.,10)
   50 CONTINUE
      RETURN
      END


      SUBROUTINE SLOPE (N,DT,X,SL,SIGMA,SA)
C     COMPUTES SLOPE OF LOG(X(T)) BY LEAST SQUARES METHOD
      DIMENSION X(1), Y(100)
      XMAX=-100.0
      B1=0.0
      B2=0.0
      DO 10 I=1,N
      Y(I)=AMAX1(ABS(X(I)),0.0000001)
```

```
      Y(I)=ALOG(Y(I))
      XMAX=AMAX1(XMAX,Y(I))
      B1=B1+Y(I)
   10 B2=B2+I*Y(I)
      IF (XMAX.LT.-10.) GO TO 30
      A=0.5*DT*N*(N+1.0)
      C=DT*DT*N*(N+1.0)*(2.0*N+1.0)/6.0
      B2=DT*B2
      DEL=N*C-A*A
      SA=(B1*C-A*B2)/DEL
      SL=(N*B2-A*B1)/DEL
      SIGMA=0.0
      DO 20 I=1,N
      DEV=Y(I)-SA-SL*DT*I
   20 SIGMA=SIGMA+DEV*DEV
      SIGMA=SQRT(SIGMA/FLOAT(N))
      GO TO 40
   30 SL=0.0
      SA=0.0
      SIGMA=0.0
   40 CONTINUE
      RETURN
      END



      SUBROUTINE FPRINT (ITER)
C     FINAL OUTPUT
      COMMON RO(10,30,30),AL(10,30,30),XO(10,30,30),XL(10,30,30),R(30,30
     1),Z(30,30),RU(30,30),ZU(30,30),RV(30,30),ZV(30,30),X(30,30),RA(30)
     2,ZA(30),RN(30),ZN(30),RB1(30),RB2(30),ZB1(30),ZB2(30),RBU1(30),RBU
     32(30),ZBU1(30),ZBU2(30),RBV1(30),RBV2(30),ZBV1(30),ZBV2(30),HB1(30
     4),HB2(30)
      COMMON /AUX/ SP(100),SD(100),SI(100),SJ(100),SIV(50),SJV(50),SPP(5
     10),SVV(50),RP(50),RM(50),RP1(50),RV1(50),F(100),FP(100),FPP(100),F
     2PPP(100),XOR(4),YOR(4),YB(4,4),SLRV1(50),SIRV1(50),SARV1(50),SLRV2
     3(50),SIRV2(50),SARV2(50),SLZV1(50),SIZV1(50),SAZV1(50),SLZV2(50),S
     4IZV2(50),SAZV2(50),SLRA1(50),SIRA1(50),SARA1(50),SLRA2(50),SIRA2(5
     50),SARA2(50),SLZA1(50),SIZA1(50),SAZA1(50),SLZA2(50),SIZA2(50),SAZ
     6A2(50),SLMK1(50),SIMK1(50),SAMK1(50),SLMK2(50),SIMK2(50),SAMK2(50)
     7,SLMK3(50),SIMK3(50),SAMK3(50),SLMK4(50),SIMK4(50),SAMK4(50),NITER
     8(50)
      COMMON /POT/ RVA(30),ZVA(30),BPV(10),BTV(10),PT(10,30,30),PP(10,30
     1,30)
      COMMON /INP/ Q(10),QT(10),AM(10),PR(10),QQ(10),PC(10),TC(10),BPP(1
     10),BTP(10),BET(10)
      COMMON /CPL/ NI,NJ,NK,EP,ZLE,GAM,SM,N1,N2,N3,N4,N5,NVAC,HS,HU,HV,P
     1I,RX,RY,E4,A1,A2,A3,A4,A5,A6,HU4,HV4,HUV,IC,IO,SA1,SA2,SA3,SE1,SE2
     2,SE3,DT,RA1,RA2,RA3,RE1,RE2,RE3,ENER,FAXIS,DG1,DG2,DG3,DH1,DH2,DH3
     3,NPLOT,I1
      COMMON /CVA/ NIV,FV1,FV2,NV,NP,OM,PM,HR,H1,H2,H3,C1,C2,N6,EVAC,ETO
     1T,A11,A22,A12,HR4,DTB,FAXV,OM1,OM2,PM1,PM2
      COMMON /FOU/ SV(7,30),SU(7,30),SFI(7,30),SRO(7,7),XR(7),XZ(7),YR(7
     1),YZ(7)
      COMMON /PLOT/ NRA1,NRA2,NZA1,NZA2,MK1,MK2,MK3,MK4,M1,M2,M3,M4,K1,K
```

```
      12,K3,K4,RNAME(7),ZNAME(7)
      ZERVAL(XE,YE,ZE,UE,VE,WE)=(UE*YE*ZE*(ZE-YE)+VE*XE*ZE*(XE-ZE)+WE*XE
     1*YE*(YE-XE))/(XE*YE*(YE-XE)+YE*ZE*(ZE-YE)+XE*ZE*(ZE-XE))
      VALZD(XE,YE,UE,VE)=(UE*YE*YE-VE*XE*XE)/(YE*YE-XE*XE)
C     CHOICES FOR NRA AND NZA FROM INPUT CARD NO.36
      RNAME(1)=10HR,CONST
      RNAME(2)=10HR,SIN(V)
      RNAME(3)=10HR,COS(V)
      RNAME(4)=10HR,SIN(2V)
      RNAME(5)=10HR,COS(2V)
      RNAME(6)=10HR,SIN(3V)
      RNAME(7)=10HR,COS(3V)
      ZNAME(1)=10HZ,CONST
      ZNAME(2)=10HZ,SIN(V)
      ZNAME(3)=10HZ,COS(V)
      ZNAME(4)=10HZ,SIN(2V)
      ZNAME(5)=10HZ,COS(2V)
      ZNAME(6)=10HZ,SIN(3V)
      ZNAME(7)=10HZ,COS(3V)
C     PRINT OUT RESULTS FOR LAST TIME STEP
      DT1=DT*IC
      PRINT 340, ENER
      IF (NVAC.LT.0) GO TO 10
      PRINT 350, EVAC,ETOT
   10 PRINT 360
      PRINT 370
      PRINT 380, (XP(I),I=1,7)
      PRINT 390, (XZ(I),I=1,7)
      IF (NVAC.LT.0) GO TO 20
      PRINT 400
      PRINT 370
      PRINT 380, (YR(I),I=1,7)
      PRINT 390, (YZ(I),I=1,7)
      PRINT 410
      GO TO 30
   20 PRINT 420, I1
   30 PRINT 370
      PRINT 430, (SRO(1,L),L=1,7)
      PRINT 440, (SRO(2,L),L=1,7)
      PRINT 450, (SRO(3,L),L=1,7)
      PRINT 460, (SRO(4,L),L=1,7)
      PRINT 470, (SRO(5,L),L=1,7)
      PRINT 480, (SRO(6,L),L=1,7)
      PRINT 490, (SRO(7,L),L=1,7)
C     COMPUTE AVERAGE RADIUS FOR EACH FLUX SURFACE
      DO 50 I=1,NI
      RP(I)=0.0
      DO 40 J=2,N1
      DO 40 K=2,N4
   40 RP(I)=RP(I)+RO(I,J,K)
      RP(I)=RP(I)*HU*HV
      IF (NVAC.GT.0) RP(I)=RP(I)*SRO(1,1)
   50 CONTINUE
      IF (NVAC.LT.0) GO TO 70
      DO 60 I=1,NIV
      DS=(I-1)*HR
   60 RM(I)=SRO(1,1)+DS*(1.0-SRO(1,1))
```

```
   70 PRINT 500
      PRINT 510
C     COMPUTE RADIUS AND MU AT MIDPOINTS IN S
      DO 80 I=1,N3
      QQ(I)=0.5*(Q(I)/QT(I)+Q(I+1)/QT(I+1))
   80 RP1(I)=0.5*(RP(I)+RP(I+1))
      RP1(1)=RP1(1)*SQRT(2.)
C     COMPUTE VALUES AT S=0
      BETO=VALZD(RP1(1),RP1(2),BET(1),BET(2))
      PRO=VALZD(RP1(1),RP1(2),PR(1),PR(2))
      QQO=ZERVAL(RP1(1),RP1(2),RP1(3),QQ(1),QQ(2),QQ(3))
      BPPO=0.
      TCO=0.
      BTPO=VALZD(RP1(1),RP1(2),BTP(1),BTP(2))
      PCO=VALZD(RP1(1),RP1(2),PC(1),PC(2))
      DO 90 J=1,N3
      RP1(NI+1-J)=RP1(NI-J)
      BTP(NI+1-J)=BTP(NI-J)
      BPP(NI+1-J)=BPP(NI-J)
      TC(NI+1-J)=TC(NI-J)
      PC(NI+1-J)=PC(NI-J)
      QQ(NI+1-J)=QQ(NI-J)
      PR(NI+1-J)=PR(NI-J)
   90 BET(NI+1-J)=BET(NI-J)
      RP1(1)=0.
      BTP(1)=BTPO
      PC(1)=PCO
      BPP(1)=0.
      TC(1)=0.
      QQ(1)=QQO
      PR(1)=PRO
      BET(1)=BETO
C     PRINT AVERAGE VALUES
      DO 100 I=1,NI
  100 PRINT 520, RP1(I),BTP(I),BPP(I),TC(I),PC(I),QQ(I),PR(I),BET(I)
      IF (NVAC.LT.0) GO TO 130
      PRINT 530
      PRINT 540, C1,C2
      PRINT 550
      DO 110 I=1,N6
  110 RV1(I)=0.5*(RM(I)+RM(I+1))
      DO 120 I=1,N6
  120 PRINT 560, RV1(I),BTV(I),BPV(I)
  130 CONTINUE
      REWIND 3
      REWIND 4
      XY=FLOAT(ITER)/FLOAT(IC)
      N=INT(XY-1.0)
      NEV=50
      NEV1=MINO(NEV,N)
      M1=MK1/10
      M2=MK2/10
      M3=MK3/10
      M4=MK4/10
      K1=MK1-M1*10
      K2=MK2-M2*10
      K3=MK3-M3*10
```

```
        K4=MK4-M4*10
        XY=FLOAT(N)/50.0
        NN=INT(XY)+1
        IF (NVAC.LT.0) GO TO 160
        NTER=0
        PRINT 570
        PRINT 580
        PRINT 590
        J=0
        JN=0
        DO 150 I=1,N
        NTER=NTER+1
        READ (4) ERBO,ERBO1,ERVA
        READ (4) (YR(M),M=1,7),(YZ(M),M=1,7)
        READ (4) OM1,OM2
        J=J+1
        F(J)=YR(NRA1)
        FP(J)=YR(NRA2)
        FPP(J)=YZ(NZA1)
        FPPP(J)=YZ(NZA2)
        IF (J.LT.NEV1) GO TO 140
C       COMPUTE GROWTH RATES FOR VACUUM AXIS
        J=0
        JN=JN+1
        CALL SLOPE (NEV1,DT1,F,SLRV1(JN),SIRV1(JN),SARV1(JN))
        CALL SLOPE (NEV1,DT1,FP,SLRV2(JN),SIRV2(JN),SARV2(JN))
        CALL SLOPE (NEV1,DT1,FPP,SLZV1(JN),SIZV1(JN),SAZV1(JN))
        CALL SLOPE (NEV1,DT1,FPPP,SLZV2(JN),SIZV2(JN),SAZV2(JN))
    140 IF (NTER.LT.NN) GO TO 150
        NTER=0
        TT=(I-1)*DT*IC
C       PRINT FOURIER COEFFICIENTS FOR VACUUM AXIS
        PRINT 600, TT,(YR(M),M=1,7),(YZ(M),M=1,7)
    150 CONTINUE
        REWIND 4
    160 NTER=0
        PRINT 610
        PRINT 580
        PRINT 590
        JJ=0
        JN=0
        DO 180 I=1,N
        NTER=NTER+1
        READ (3) ETOT,ERO,EAL,EAX
        READ (3) (XR(M),M=1,7),(XZ(M),M=1,7)
        READ (3) ((YB(J,K),K=1,4),J=1,4)
        READ (3) EA1,EA2,EA3
        JJ=JJ+1
        F(JJ)=XR(NRA1)
        FP(JJ)=XR(NRA2)
        FPP(JJ)=XZ(NZA1)
        FPPP(JJ)=XZ(NZA2)
        IF (JJ.LT.NEV1) GO TO 170
C       COMPUTE GROWTH RATES FOR MAGNETIC AXIS
        JJ=0
        JN=JN+1
        NITER(JN)=(I-1)*IC
```

```
      CALL SLOPE (NEV1,DT1,F,SLRA1(JN),SIRA1(JN),SARA1(JN))
      CALL SLOPE (NEV1,DT1,FP,SLRA2(JN),SIRA2(JN),SARA2(JN))
      CALL SLOPE (NEV1,DT1,FPP,SLZA1(JN),SIZA1(JN),SAZA1(JN))
      CALL SLOPE (NEV1,DT1,FPPP,SLZA2(JM),SIZA2(JN),SAZA2(JN))
  170 IF (NTER.LT.NN) GO TO 180
      NTER=0
      TT=(I-1)*DT*IC
C     PRINT FOURIER COEFFICIENTS FOR MAGNETIC AXIS
      PRINT 600, TT,(XR(M),M=1,7),(XZ(M),M=1,7)
  180 CONTINUE
      NTER=0
      REWIND 3
      IF (NVAC.LT.0) GO TO 190
      PRINT 620
      GO TO 200
  190 PRINT 630, I1
  200 PRINT 640
      JJ=0
      JN=0
      DO 220 I=1,N
      NTER=NTER+1
      READ (3) ETOT,ERO,EAL,EAX
      READ (3) (XR(M),M=1,7),(XZ(M),M=1,7)
      READ (3) ((YB(J,K),J=1,4),K=1,4)
      READ (3) EA1,EA2,EA3
      JJ=JJ+1
      F(JJ)=YB(M1+1,K1+1)
      FP(JJ)=YB(M2+1,K2+1)
      FPP(JJ)=YB(M3+1,K3+1)
      FPPP(JJ)=YB(M4+1,K4+1)
      IF (JJ.LT.NEV1) GO TO 210
C     COMPUTE GROWTH RATES FOR FLUX SURFACE
      JJ=0
      JN=JN+1
      CALL SLOPE (NEV1,DT1,F,SLMK1(JN),SIMK1(JN),SAMK1(JN))
      CALL SLOPE (NEV1,DT1,FP,SLMK2(JN),SIMK2(JN),SAMK2(JN))
      CALL SLOPE (NEV1,DT1,FPP,SLMK3(JN),SIMK3(JN),SAMK3(JN))
      CALL SLOPE (NEV1,DT1,FPPP,SLMK4(JN),SIMK4(JN),SAMK4(JN))
  210 IF (NTER.LT.NN) GO TO 220
      NTER=0
      TT=(I-1)*DT*IC
C     PRINT FOURIER COEFFICIENTS FOR FLUX SURFACE
      PRINT 650, TT,((YB(J,K),J=1,4),K=1,4)
  220 CONTINUE
      REWIND 3
      REWIND 4
      NTER=0
      IF (NVAC.GT.0) GO TO 230
      PRINT 660
      PRINT 670
      GO TO 240
  230 PRINT 680
      PRINT 690
  240 DO 270 I=1,N
      NTER=NTER+1
      READ (3) ETOT,ERO,EAL,EAX
      READ (3) (XR(M),M=1,7),(XZ(M),M=1,7)
```

```
      READ (3) ((YB(J,K),K=1,4),J=1,4)
      READ (3) EA1,EA2,EA3
      IF (NVAC.LT.0) GO TO 250
      READ (4) ERB0,ERB01,ERVA
      READ (4) (YR(M),M=1,7),(YZ(M),M=1,7)
      READ (4) OM1,OM2
  250 IF (NTER.LT.NN) GO TO 270
C     PRINT DESCENT AND RELAXATION COEFFICIENTS
      NTER=0
      TT=(I-1)*DT*IC
      IF (NVAC.LT.0) GO TO 260
      PRINT 700, TT,EA1,EA2,EA3,OM1,OM2
      GO TO 270
  260 PRINT 710, TT,EA1,EA2,EA3
  270 CONTINUE
      IF (JN.LT.1) GO TO 330
C     PRINT GROWTH RATES
      PRINT 720
      PRINT 730
      PRINT 740, RNAME(NRA1),RNAME(NRA2),ZNAME(NZA1),ZNAME(NZA2)
      PRINT 750
      DO 280 J=1,JN
  280 PRINT 760, NITER(J),SLRA1(J),SIRA1(J),SARA1(J),SLRA2(J),SIRA2(J),S
     1ARA2(J),SLZA1(J),SIZA1(J),SAZA1(J),SLZA2(J),SIZA2(J),SAZA2(J)
      IF (NVAC.LT.0) GO TO 300
      PRINT 770
      PRINT 740, RNAME(NRA1),RNAME(NRA2),ZNAME(NZA1),ZNAME(NZA2)
      PRINT 750
      DO 290 J=1,JN
  290 PRINT 760, NITER(J),SLRV1(J),SIRV1(J),SARV1(J),SLRV2(J),SIRV2(J),S
     1ARV2(J),SLZV1(J),SIZV1(J),SAZV1(J),SLZV2(J),SIZV2(J),SAZV2(J)
      PRINT 780
      GO TO 310
  300 PRINT 790, I1
  310 PRINT 800, M1,K1,M2,K2,M3,K3,M4,K4
      PRINT 750
      DO 320 J=1,JN
  320 PRINT 760, NITER(J),SLMK1(J),SIMK1(J),SAMK1(J),SLMK2(J),SIMK2(J),S
     1AMK2(J),SLMK3(J),SIMK3(J),SAMK3(J),SLMK4(J),SIMK4(J),SAMK4(J)
  330 CONTINUE
      RETURN
  340 FORMAT (1H1///6X10H3D RESULTS///6X14HPLASMA ENERGY=E16.9)
  350 FORMAT (/6X14HVACUUM ENERGY=E16.9//6X14HTOTAL ENERGY =E16.9)
  360 FORMAT (////6X34HFOURIER COEFFICIENTS MAGNETIC AXIS//)
  370 FORMAT (25X5HCONST,5X6HSIN(V),5X6HCOS(V),4X7HSIN(2V),4X7HCOS(2V),1
     17X7HSIN(3V),4X7HCOS(3V)//)
  380 FORMAT (16X1HR2X5F11.5,13X2F11.5/)
  390 FORMAT (16X1HZ,2X5F11.5,13X2F11.5/)
  400 FORMAT (//6X32HFOURIER COEFFICIENTS VACUUM AXIS//)
  410 FORMAT (//6X34HFOURIER COEFFICIENTS FREE BOUNDARY//)
  420 FORMAT (////6X39HFOURIER COEFFICIENTS FLUX SURFACE AT I=I3//)
  430 FORMAT (11X6HCONST 2X5F11.5,13X2F11.5/)
  440 FORMAT (11X6HSIN(U),2X5F11.5,13X2F11.5/)
  450 FORMAT (11X6HCOS(U),2X5F11.5,13X2F11.5/)
  460 FORMAT (11X7HSIN(2U)1X5F11.5,13X2F11.5/)
  470 FORMAT (11X7HCOS(2U)1X5F11.5,13X2F11.5/)
  480 FORMAT (11X7HSIN(3U)1X5F11.5,13X2F11.5/)
```

```
  490 FORMAT (11X7HCOS(3U)1X5F11.5,13X2F11.5/)
  500 FORMAT (1H1///6X13HPLASMA REGION//)
  510 FORMAT (8X6HRADIUS,5X4HBTOR,5X4HBPOL,4X5HTCURR,4X5HPCURR,6X2HMU8X1
     1HP,6X4HBETA/)
  520 FORMAT (5X,8F9.3)
  530 FORMAT (/////6X13HVACUUM REGION//)
  540 FORMAT (8X17HTOROIDAL CURRENT=F8.3,3X17HPOLOIDAL CURRENT=F8.3)
  550 FORMAT (//8X6HRADIUS,5X4HBTOR,5X4HBPOL/)
  560 FORMAT (5X3F9.3)
  570 FORMAT (1H1,50X32HFOURIER COEFFICIENTS VACUUM AXIS/)
  580 FORMAT (39X,1HR,55X1HZ/)
  590 FORMAT (7X4HTIME,3X5HCONST,3X5HSI(V),3X5HCO(V),2X6HSI(2V),2X6HCO(2
     1V)2X6HSI(3V),2X6HCO(3V),3X5HCONST,3X5HSI(V),3X5HCO(V),2X6HSI(2V),2
     2X6HCO(2V),2X6HSI(3V),2X6HCO(3V)/)
  600 FORMAT (5X,F6.2,14F8.4)
  610 FORMAT (1H1,50X34HFOURIER COEFFICIENTS MAGNETIC AXIS/)
  620 FORMAT (1H1,40X34HFOURIER COEFFICIENTS FREE BOUNDARY/)
  630 FORMAT (1H1,40X33HFOURIER COEFFICIENTS RADIUS AT I=I3/)
  640 FORMAT (7X4HTIME,2X5HMK=00,2X5HMK=10,2X5HMK=20,2X5HMK=30,2X5HMK=01
     1,2X5HMK=11,2X5HMK=21,2X5HMK=31,2X5HMK=02,2X5HMK=12,2X5HMK=22,2X5HM
     2K=32,2X5HMK=03,2X5HMK=13,2X5HMK=23,2X5HMK=33/)
  650 FORMAT (5XF6.2,16F7.3)
  660 FORMAT (1H1///17X20HDESCENT COEFFICIENTS/)
  670 FORMAT (8X4HTIME,4X5HE1/A1,3X5HE2/A2,3X5HE3/A3/)
  680 FORMAT (1H1///17X20HDESCENT COEFFICIENTS,16X18HRELAXATION FACTORS/
     1)
  690 FORMAT (8X4HTIME,4X5HE1/A1,3X5HE2/A2,3X5HE3/A3,19X3HOM1,5X3HOM2/)
  700 FORMAT (6X,F6.2,2X3F8.3,14X2F8.3)
  710 FORMAT (6X,F6.2,2X3F8.3)
  720 FORMAT (1H1///6X19HGROWTH RATES LAMBDA/12X5HWHERE/18X25HLOG(ABS(F(
     1T)))=LAMBDA*T+A/18X31HSIGMA IS THE STANDARD DEVIATION//)
  730 FORMAT (//64X,13HMAGNETIC AXIS/)
  740 FORMAT (19XA10,22XA10,22XA10,22X,A10/)
  750 FORMAT (6X4HITER,4X6HLAMBDA3X5HSIGMA,5X1HA11X6HLAMBDA,3X5HSIGMA,5X
     11HA,11X6HLAMBDA,3X5HSIGMA,5X1HA,11X6HLAMBDA,3X5HSIGMA,5X1HA/)
  760 FORMAT (6XI4,2X3F8.3,3(7X,3(F8.3)))
  770 FORMAT (//64X11HVACUUM AXIS/)
  780 FORMAT (//64X13HFREE BOUNDARY/)
  790 FORMAT (//64X12HRADIUS AT I=I2/)
  800 FORMAT (22X3HMK=2I1,26X3HMK=2I1,26X3HMK=2I1,26X3HMK=2I1/)
      END



      SUBROUTINE SPLIF (M,N,S,F,FP,FPP,FPPP,KM,VM,KN,VN,MODE,FQM,IND)
C     SPLINE FIT-ANTONY JAMESON
      DIMENSION S(1), F(1), FP(1), FPP(1), FFPP(1)
      IND=0
      K=IABS(N-M)
      IF (K-1) 180,180,10
   10 K=(N-M)/K
      I=M
      J=M+K
      DS=S(J)-S(I)
      D=DS
```

```
      IF (DS) 20,180,20
   20 DF=(F(J)-F(I))/DS
      IF (KM-2) 30,40,50
   30 U=.5
      V=3.*(DF-VM)/DS
      GO TO 80
   40 U=0.
      V=VM
      GO TO 80
   50 U=-1.
      V=-DS*VM
      GO TO 80
   60 I=J
      J=J+K
      DS=S(J)-S(I)
      IF (D*DS) 180,180,70
   70 DF=(F(J)-F(I))/DS
      B=1./(DS+DS+U)
      U=B*DS
      V=B*(6.*DF-V)
   80 FP(I)=U
      FPP(I)=V
      U=(2.-U)*DS
      V=6.*DF+DS*V
      IF (J-N) 60,90,60
   90 IF (KN-2) 100,110,120
  100 V=(6.*VN-V)/U
      GO TO 130
  110 V=VN
      GO TO 130
  120 V=(DS*VN+FPP(I))/(1.+FP(I))
  130 B=V
      D=DS
  140 DS=S(J)-S(I)
      U=FPP(I)-FP(I)*V
      FPPP(I)=(V-U)/DS
      FPP(I)=U
      FP(I)=(F(J)-F(I))/DS-DS*(V+U+U)/6.
      V=U
      J=I
      I=I-K
      IF (J-M) 140,150,140
  150 I=N-K
      FPPP(N)=FPPP(I)
      FPP(N)=B
      FP(N)=DF+D*(FPP(I)+B+B)/6.
      IND=1
      IF (MODE) 180,180,160
  160 FPPP(J)=FQM
      V=FPP(J)
  170 I=J
      J=J+K
      DS=S(J)-S(I)
      U=FPP(J)
      FPPP(J)=FPPP(I)+.5*DS*(F(I)+F(J)-DS*DS*(U+V)/12.)
      V=U
      IF (J-N) 170,180,170
```

```
  180 RETURN
      END



      SUBROUTINE INTPL (MI,NI,SI,FI,M,N,S,F,FP,FPP,FPPP,MODE)
C     INTERPOLATION USING TAYLOR SERIES-ANTONY JAMESON
      DIMENSION SI(1), FI(1), S(1), F(1), FP(1), FPP(1), FPPP(1)
      K=IABS(N-M)
      K=(N-M)/K
      I=M
      MIN=MI
      NIN=NI
      D=S(N)-S(M)
      IF (D*(SI(NI)-SI(MI))) 10,20,20
   10 MIN=NI
      NIN=MI
   20 KI=IABS(NIN-MIN)
      IF (KI) 40,40,30
   30 KI=(NIN-MIN)/KI
   40 II=MIN-KI
      C=0.
      IF (MODE) 60,60,50
   50 C=1.
   60 II=II+KI
      SS=SI(II)
   70 I=I+K
      IF (I-N) 80,90,80
   80 IF (D*(S(I)-SS)) 70,70,90
   90 J=I
      I=I-K
      SS=SS-S(I)
      FPPPP=C*(FPPP(J)-FPPP(I))/(S(J)-S(I))
      FF=FPPP(I)+.25*SS*FPPPP
      FF=FPP(I)+SS*FF/3.
      FF=FP(I)+.5*SS*FF
      FI(II)=F(I)+SS*FF
      IF (II-NIN) 60,100,60
  100 RETURN
      END



      SUBROUTINE ASYM (ASYE)
C     MAIN AXIALLY SYMMETRIC PROGRAM
      COMMON /AUX/ R2,Z2,RA,ZA,X(30),R(10,30),AL(10,30),PA(10,30),D(10,3
     10),E(10,30),F(10,30),G(10,30),R1(10,30),Z1(10,30),GR(10,30),GZ(10,
     230),DD(10,30),QR(10,30),QZ(10,30),A(30),B(30),C(30),H(30),RR(30),Z
     33Z(30),RU(30),ZU(30),RB(30),ZB(30),RBU(30),ZBU(30),XA(30),XC(30),XP
     4D(30),XPK(30),AK2(30),AKF(30),AKFU(30),GAXR,GAXZ
      COMMON /INP/ Q(10),QT(10),AM(10),PR(10),QQ(10),PC(10),TC(10),BPP(1
     10),BTP(10),BET(10)
      COMMON /AC/ EN(100,5),ET(5),AA1(5),AA2(5),AA3(5),AVE(5),NAC,NE,SAF
```

```
      1I,SAFV,SAFPSI,SAFRO,SAFAX
       COMMON /CPL/ NI,NJ,NK,EP,ZLE,GAM,SM,N1,N2,N3,N4,N5,NVAC,HS,HU,HV,P
      1I,RX,RY,E4,A1,A2,A3,A4,A5,A6,HU4,HV4,HUV,IC,IO,SA1,SA2,SA3,SE1,SE2
      2,SE3,DT,RA1,RA2,RA3,RE1,RE2,PE3,ENER,FAXIS,DG1,DG2,DG3,DH1,DH2,DH3
      3,NPLOT,I1
       COMMON /CVA/ NIV,FV1,FV2,NV,NP,OM,PM,HR,H1,H2,H3,C1,C2,N6,EVAC,ETO
      1T,A11,A22,A12,HR4,DTB,FAXV,OM1,OM2,PM1,PM2
       DO 10 J=1,4
   10 AVE(J)=0.0
       NJA=4
       IF (NVAC.LT.0) NJA=3
C      INITIAL VALUES FOR DESCENT COEFFICIENTS
       DO 20 I=1,NE
       EN(I,1)=SE1/SAFRO
       EN(I,2)=SE2/SAFPSI
       EN(I,3)=SE3/SAFAX
   20 CONTINUE
       IF (NVAC.LT.0) GO TO 40
       DO 30 I=1,NE
   30 EN(I,4)=(2.0-OM)/(OM*DT*SAFV)
       OM1=OM
       PM1=1.0-OM
   40 CONTINUE
       DT2=2.0*DT
       DO 50 J=1,4
   50 AA2(J)=0.0
       INI=0
       ELIM=2.0
       ERBO=0.0
       ERBO1=0.0
       EVNE=0.0
       ERVA=0.0
       C1=1.0
       NE1=NE-1
       HU2=2.0*HU
       IO1=2*IO
       INI=0
       IT=0
       ITER=0
       IC1=2*IC
       NTER=1000
       KTER=1000
       ENER=0.0
       EVAC=0.0
       ETOT=0.0
C      EVALUATES FLUX AND MASS FUNCTIONS AND INITIALIZES AXIALLY
C      SYMMETRIC SOLUTION
       CALL ASIN
       DO 60 I=1,NI
       DO 60 J=1,N2
       R1(I,J)=R(I,J)
   60 Z1(I,J)=AL(I,J)
       RAX=RA
       ZAX=ZA
   70 CONTINUE
       DG1=1.0/(RA1+RE1)
       DG2=1.0/(RA2+RE2)
```

```
      DG3=1.0/(RA3+RE3)
      DH1=2.0*RA1+RE1
      DH2=2.0*RA2+RE2
      DH3=2.0*RA3+RE3
C     COMPUTE BOUNDARY QUANTITIES
      DO 80 J=1,N1
      X1=0.5*(X(J)+X(J+1))
      X2=(X(J+1)-X(J))/HU
      RB(J)=R2+X1*(RR(J)-R2)
      ZB(J)=Z2+X1*(ZZ(J)-Z2)
      RBU(J)=X2*(RR(J)-R2)+X1*RU(J)
   80 ZBU(J)=X2*(ZZ(J)-Z2)+X1*ZU(J)
      IF (NVAC.LT.0) GO TO 90
      NCO=0
C     COMPUTES NV ITERATIONS OF VACUUM EQUATIONS
      CALL ASOR (NCO,ERVA,EVNE,IT)
   90 NCO=0
  100 CONTINUE
C     COMPUTES SPACE OPERATORS FOR PLASMA EQUATIONS
      CALL GRAD (ENEW,RMAX,ALMA,AXIE,SJA,BJA)
C     MAGNETIC AXIS ITERATION
      PAS1=DG3*(RA*DH3-GAXR-RA3*RAX)
      PAS2=DG3*(ZA*DH3-GAXZ-RA3*ZAX)
      RAX=RA
      ZAX=ZA
      RA=PAS1
      ZA=PAS2
C     R EQUATION ITERATION
      DO 110 J=2,N1
      DO 110 I=2,N3
      PAS=DG1*(R(I,J)*DH1-GR(I,J)-RA1*R1(I,J))
      R1(I,J)=R(I,J)
  110 R(I,J)=PAS
C     ^SI EQUATION ITERATION
      AL(1,2)=0.0
      SUM1=0.0
      DO 120 J=2,N1
  120 SUM1=SUM1+1.0/E(1,J)
      SUM1=-QT(1)/SUM1
      DO 130 J=2,N1
  130 AL(1,J+1)=AL(1,J)+SUM1/E(1,J)
      DO 140 J=2,N1
      DO 140 I=2,NI
      PAS=DG2*(AL(I,J)*DH2-GZ(I,J)-RA2*Z1(I,J))
      Z1(I,J)=AL(I,J)
  140 AL(I,J)=PAS
      DO 150 I=2,N3
      R1(I,1)=R1(I,N1)
      R(I,1)=R(I,N1)
  150 R(I,N2)=R(I,2)
      DO 160 I=1,NI
      Z1(I,1)=Z1(I,N1)+QT(I)
      AL(I,1)=AL(I,N1)+QT(I)
  160 AL(I,N2)=AL(I,2)-QT(I)
      NCO=NCO+1
      IF (NCO.LT.NP) GO TO 100
      IF (NVAC.LT.0) GO TO 170
```

```
C       COMPUTES ONE ITERATION OF FREE BOUNDARY EQUATION
        CALL ASBO (ERBO,ERBO1)
  170 CONTINUE
        ITER=ITER+1
        IF (ITER.GE.4000) STOP
        NTER=NTER+1
        IF (ITER.LT.NAC) GO TO 210
C       COMPUTE NEW DESCENT COEFFICIENTS
        DO 190 J=1,NJA
        ET(J)=(AA1(J)-AA3(J))/DT2
        DO 180 I=1,NE1
  180 EN(I,J)=EN(I+1,J)
  190 EN(NE,J)=ABS(ET(J))/AA2(J)
        DO 200 J=1,NJA
        Y1=(EN(NE,J)-EN(NE-1,J))/(DT*0.5*(EN(NE,J)+EN(NE-1,J)))
        Y2=ABS(Y1)
        IF (Y2.GT.ELIM) EN(NE,J)=EN(NE-1,J)*(1.0+SIGN(ELIM,Y1)*DT)
  200 CONTINUE
  210 CONTINUE
        DO 220 J=1,4
        AA3(J)=AA2(J)
  220 AA2(J)=AA1(J)
        IF (ITER.LT.NAC) GO TO 250
        DO 240 J=1,NJA
        AVE(J)=0.0
        DO 230 I=1,NE
  230 AVE(J)=AVE(J)+EN(I,J)
  240 AVE(J)=AVE(J)/NE
        EA1=SAFRO*AVE(1)
        EA2=SAFPSI*AVE(2)
        EA3=SAFAX*AVE(3)
        RE1=EA1/DT
        RE2=EA2/DT
        RE3=EA3/DT
        IF (NVAC.LT.0) GO TO 250
        OM1=2.0/(1.0+SAFV*AVE(4)*DT)
        PM1=1.0-OM1
  250 CONTINUE
        KTER=KTER+1
        PMAX=AMAX1(RMAX,ALMA,AXIE,ERBO)
C       PRINTOUT AND ERROR CRITERIA
        IF (KTER.GE.IO1) GO TO 280
  260 IF (NTER.GE.IC1) GO TO 290
  270 IF (PMAX.LT.ASYE) GO TO 300
        GO TO 70
  280 KTER=0
        INI=0
        GO TO 260
  290 NTER=0
        GENP=ENEW-ENER
        GENV=EVNE-EVAC
        ENER=ENEW
        EVAC=EVNE
        ETOT=EVAC+ENER
        GEN=GENP+GENV
        IF (INI.EQ.0) PRINT 310
        INI=1
```

```
C      PRINT ITERATION DATA
       PRINT 320, ITER,SJA,BJA,GEN,AXIE,RMAX,ALMA,ERVA,ERBO,ERBO1
       GO TO 270
C      PRINT FINAL RESULTS FOR AXIALLY SYMMETRIC SOLUTION
  300 CALL ASOUT
       RETURN
  310 FORMAT (1H1///,6X4HITER,1X7HJAC MAX,1X7HJAC MIN,3X10HDEL ENERGY,2X
      110HAXIS ERROR,2X10H RHO ERROR,2X10H PSI ERROR,2X,10H VAC ERROR,2X1
      20H BOU ERROR,2X10H  VAC AXIS/)
  320 FORMAT (5X15,2F8.3,E13.4,6E12.4)
       END



       SUBROUTINE GRAD (ENEW,RMAX,ALMA,AXIE,SJA,BJA)
C      COMPUTES SPACE OPERATORS FOR PLASMA EQUATIONS
       COMMON /AUX/ R2,Z2,RA,ZA,X(30),R(10,30),AL(10,30),PA(10,30),D(10,3
      10),E(10,30),F(10,30),G(10,30),R1(10,30),Z1(10,30),GR(10,30),GZ(10,
      230),DD(10,30),QR(10,30),QZ(10,30),A(30),B(30),C(30),H(30),RR(30),Z
      3Z(30),RU(30),ZU(30),RB(30),ZB(30),RBU(30),ZBU(30),XA(30),XC(30),XP
      4D(30),XPK(30),AK2(30),AKF(30),AKFU(30),GAXR,GAXZ
       COMMON /INP/ Q(10),QT(10),AM(10),PR(10),QQ(10),PC(10),TC(10),BPP(1
      10),BTP(10),BET(10)
       COMMON /AC/ EN(100,5),ET(5),AA1(5),AA2(5),AA3(5),AVE(5),NAC,NE,SAF
      1I,SAFV,SAFPSI,SAFRO,SAFAX
       COMMON /CPL/ NI,NJ,NK,EP,ZLE,GAM,SM,N1,N2,N3,N4,N5,NVAC,HS,HU,HV,P
      1I,RX,RY,E4,A1,A2,A3,A4,A5,A6,HU4,HV4,HJV,IC,IO,SA1,SA2,SA3,SE1,SE2
      2,SE3,DT,RA1,RA2,RA3,RE1,RE2,RE3,ENER,FAXIS,DG1,DG2,DG3,DH1,DH2,DH3
      3,NPLOT,I1
       COMMON /CVA/ NIV,FV1,FV2,NV,NP,OM,PM,HR,H1,H2,H3,C1,C2,N6,EVAC,ETO
      1T,A11,A22,A12,HR4,DTB,FAXV,OM1,OM2,PM1,PM2
       DO 10 J=1,3
   10 AA1(J)=0.0
       ENEW=0.0
       RMAX=0.0
       ALMA=0.0
       GAXR=0.0
       GAXZ=0.0
       SJA=-1000.0
       BJA=1000.0
       HUU=HU*HU
       B1=1.0/(2.0*HU*HU)
       B2=1.0/(2.0*HS)
       HS4=4.0*HS
       GAM1=GAM-1.0
C      COMPUTE COEFFICIENTS FOR PLASMA EQUATIONS
       DO 20 J=1,N1
       X1=RB(J)-RA
       X2=ZB(J)-ZA
       A(J)=X1*X1+X2*X2
       B(J)=RBU(J)*RBU(J)+ZBU(J)*ZBU(J)
       C(J)=X1*RBU(J)+X2*ZBU(J)
   20 H(J)=X1*ZBU(J)-X2*RBU(J)
       DO 30 I=1,NI
       DO 30 J=2,N1
```

```
 30 F(I,J)=1.0+EP*(RA+0.5*(RB(J)-RA)*(R(I,J)+R(I,J+1)))
    DO 50 I=1,N3
    SUM1=0.0
    DO 40 J=2,N1
    G(I,J)=H(J)*(R(I+1,J)*R(I+1,J)-R(I,J)*R(I,J)+R(I+1,J+1)*R(I+1,J+1)
   1-R(I,J+1)*R(I,J+1))/HS4
    SJA=AMAX1(SJA,G(I,J))
    BJA=AMIN1(BJA,G(I,J))
 40 SUM1=SUM1+0.5*(F(I,J)+F(I+1,J))*G(I,J)
    IF (SUM1.LE.0.0) GO TO 160
 50 PR(I)=(AM(I)/(SUM1*HU))**GAM
    DO 60 J=2,N1
    DD(1,J)=1.0/G(1,J)
    DD(NI,J)=1.0/G(N3,J)
    DO 60 I=2,N3
 60 DD(I,J)=1.0/G(I,J)+1.0/G(I-1,J)
    DO 70 J=2,N1
    XA(J)=0.0
    XC(J)=0.0
    DO 70 I=1,NI
    X1=(R(I,J+1)-R(I,J))*(R(I,J+1)-R(I,J))/HUU
    X2=0.5*(R(I,J)*R(I,J)+R(I,J+1)*R(I,J+1))
    X3=(R(I,J+1)*R(I,J+1)-R(I,J)*R(I,J))/HU
    X4=(AL(I,J+1)-AL(I,J))*(AL(I,J+1)-AL(I,J))/HUU
    RU2=A(J)*X1+B(J)*X2+C(J)*X3
    E(I,J)=F(I,J)*DD(I,J)
    D(I,J)=0.5*QQ(I)*DD(I,J)/F(I,J)
    QR(I,J)=0.5*((QQ(I)*RU2)/F(I,J)+F(I,J)*X4)
    QZ(I,J)=0.5*((QQ(I)*RU2)/(F(I,J)*F(I,J))-X4)*DD(I,J)
    XA(J)=XA(J)+0.5*D(I,J)*X1
 70 XC(J)=XC(J)+0.5*D(I,J)*X3
    DO 100 J=2,N1
    XPK(J)=0.0
    DO 80 I=1,N3
    X1=(QR(I,J)+QR(I+1,J))/G(I,J)
    X2=PR(I)*(F(I,J)+F(I+1,J))
    ENEW=ENEW+0.5*(X1+X2*G(I,J)/GAM1)
    QR(I,J)=0.5*(X1/G(I,J)+X2)
 80 XPK(J)=XPK(J)+QR(I,J)*G(I,J)/H(J)
    DO 90 I=2,N3
 90 QZ(I,J)=QZ(I,J)+PR(I)*G(I,J)+PR(I-1)*G(I-1,J)
    QZ(1,J)=QZ(1,J)+PR(1)*G(1,J)
100 QZ(NI,J)=QZ(NI,J)+PR(N3)*G(N3,J)
    DO 110 J=2,N1
    XPD(J)=0.0
    E(NI,J)=2.0*E(NI,J)
    DO 110 I=1,NI
110 XPD(J)=XPD(J)+QZ(I,J)*0.5*(1.0-0.5*(R(I,J)+R(I,J+1)))*EP
    DO 120 I=1,NI
    E(I,1)=E(I,N1)
    D(I,1)=D(I,N1)
    QR(I,1)=QR(I,N1)
120 QZ(I,1)=QZ(I,N1)
    XA(1)=XA(N1)
    XC(1)=XC(N1)
    XPK(1)=XPK(N1)
    XPD(1)=XPD(N1)
```

```
      ENEW=ENEW*HS*HU*ZLE
C     COMPUTE MAGNETIC AXIS OPERATORS
      DO 130 J=2,N1
      GAXR=GAXR-XPD(J)+XPK(J)*ZBU(J)-2.0*XA(J)*(RB(J)-RA)-XC(J)*RBU(J)
  130 GAXZ=GAXZ-XPK(J)*RBU(J)-2.0*XA(J)*(ZB(J)-ZA)-XC(J)*ZBU(J)
      GAXR=GAXR*HS*HU
      GAXZ=GAXZ*HS*HU
      AA=ABS(GAXR)
      BB=ABS(GAXZ)
      GAXR=GAXR*FAXIS
      GAXZ=GAXZ*FAXIS
      AXIE=AMAX1(AA,BB)
C     COMPUTE OPERATOR FOR R EQUATION
      DO 140 I=2,N3
      DO 140 J=2,N1
      GR(I,J)=-E4*(QZ(I,J)*(RB(J)-RA)+QZ(I,J-1)*(RB(J-1)-RA))+R(I,J)*(B2
     1*(H(J)*(QR(I,J)-QR(I-1,J))+H(J-1)*(QR(I,J-1)-QR(I-1,J-1)))-(D(I,J)
     2*C(J)-D(I,J-1)*C(J-1))/HU+0.5*(D(I,J)*B(J)+D(I,J-1)*B(J-1)))-(D(I,
     3J)*A(J)*(R(I,J+1)-R(I,J))-D(I,J-1)*A(J-1)*(R(I,J)-R(I,J-1)))/HUU
      AA1(1)=AA1(1)+GR(I,J)*GR(I,J)
      AA=ABS(GR(I,J))
  140 RMAX=AMAX1(RMAX,AA)
C     COMPUTE OPERATOR FOR PSI EQUATION
      DO 150 I=2,NI
      DO 150 J=2,N1
      GZ(I,J)=-(E(I,J)*(AL(I,J+1)-AL(I,J))-E(I,J-1)*(AL(I,J)-AL(I,J-1)))
     1*B1
      AA=ABS(GZ(I,J))
      ALMA=AMAX1(AA,ALMA)
  150 AA1(2)=AA1(2)+GZ(I,J)*GZ(I,J)
      AA1(3)=GAXR*GAXR+GAXZ*GAXZ
      AA1(2)=AA1(2)*HU*HS
      AA1(1)=AA1(1)*HU*HS
      GO TO 170
  160 PRINT 180
      PRINT 190, ((G(I,J),I=1,N3),J=2,N1)
      CALL ASOUT
      STOP
  170 RETURN
  180 FORMAT (1H1,6X17HNEGATIVE JACOBIAN//)
  190 FORMAT (2X,11F9.3)
      END



      SUBROUTINE ASOUT
C     PRINTS FINAL RESULTS FOR AXIALLY SYMMETRIC SOLUTION
      COMMON /AUX/ R2,Z2,RA,ZA,X(30),R(10,30),AL(10,30),PA(10,30),D(10,3
     10),E(10,30),F(10,30),G(10,30),R1(10,30),Z1(10,30),GR(10,30),GZ(10,
     230),DD(10,30),QR(10,30),QZ(10,30),A(30),B(30),C(30),H(30),RR(30),Z
     3Z(30),RU(30),ZU(30),RB(30),ZB(30),RBU(30),ZBU(30),XA(30),XC(30),XP
     4D(30),XPK(30),AK2(30),AKF(30),AKFU(30),GAXR,GAXZ
      COMMON /FOU/ SV(7,30),SU(7,30),SFI(7,30),SRO(7,7),XR(7),XZ(7),YR(7
     1),YZ(7)
      COMMON /PLOT/ NRA1,NRA2,NZA1,NZA2,MK1,MK2,MK3,MK4,M1,M2,M3,M4,K1,K
```

```
   12,K3,K4,RNAME(7),ZNAME(7)
    COMMON /INP/ Q(10),QT(10),AM(10),PR(10),QQ(10),PC(10),TC(10),BPP(1
   10),BTP(10),BET(10)
    COMMON /CPL/ NI,NJ,NK,EP,ZLE,GAM,SM,N1,N2,N3,N4,N5,NVAC,HS,HU,HV,P
   1I,RX,RY,E4,A1,A2,A3,A4,A5,A6,HU4,HV4,HUV,IC,IO,SA1,SA2,SA3,SE1,SE2
   2,SE3,DT,RA1,RA2,RA3,RE1,KE2,RE3,ENER,FAXIS,DG1,DG2,DG3,DH1,DH2,DH3
   3,NPLOT,I1
    COMMON /CVA/ NIV,FV1,FV2,NV,NP,OM,PM,HR,H1,H2,H3,C1,C2,N6,EVAC,ETO
   1T,A11,A22,A12,HR4,DTB,FAXV,OM1,OM2,PM1,PM2
    PRINT 40
    PRINT 50, RA,ZA,ENER
    PRINT 60
    PRINT 70, (PR(I),I=1,N3)
    IF (NVAC.LT.0) GO TO 30
    DO 20 I=1,7
    XR(I)=0.0
    DO 10 J=2,N1
 10 XR(I)=XR(I)+X(J)*SU(I,J)
 20 XR(I)=2.0*HU*XR(I)
    XR(1)=0.5*XR(1)
    PRINT 80, R2,Z2,EVAC,ETOT
    PRINT 90, C1,C2,A11,A22
    PRINT 100
    PRINT 110, (XR(I),I=1,7)
 30 CONTINUE
    RETURN
 40 FORMAT (1H1///6X10H2D RESULTS///)
 50 FORMAT (6X13HMAGNETIC AXIS,10X2HR=F7.3,4X2HZ=F7.3//17X14HPLASMA EN
   1ERGY=E15.8//)
 60 FORMAT (6X,8HPRESSURE//)
 70 FORMAT (6X,8F7.3/8F7.3)
 80 FORMAT (////6X11HVACUUM AXIS12X2HR=F7.3,4X2HZ=F7.3//17X14HVACUUM EN
   1ERGY=E15.8,3X13HTOTAL ENERGY=E15.8//)
 90 FORMAT (/14X17HTOROIDAL CURRENT=F8.4,6X,17HPOLOIDAL CURRENT=F8.4/,
   127X4HA11=F8.4,19X4HA22=F8.4//)
100 FORMAT (///6X38HFOURIER COEFFICIENTS FOR FREE BOUNDARY///)
110 FORMAT (14X5HCONST,3X6HSIN(U),3X6HCOS(U),2X7HSIN(2U),2X7HCOS(2U)2X
   17HSIN(3U),2X7HCOS(3U)/10X,7(2X,F7.4))
    END



    SUBROUTINE ASOR (NCO,ERVA,EVNE,IT)
C   COMPUTES NV ITERATIONS FOR VACUUM EQUATIONS
    COMMON /AUX/ R2,Z2,RA,ZA,X(30),R(10,30),AL(10,30),PA(10,30),D(10,3
   10),E(10,30),F(10,30),G(10,30),R1(10,30),Z1(10,30),GR(10,30),GZ(10,
   230),DD(10,30),QR(10,30),QZ(10,30),A(30),B(30),C(30),H(30),RR(30),Z
   3Z(30),RU(30),ZU(30),RB(30),ZB(30),RBU(30),ZBU(30),XA(30),XC(30),XP
   40(30),XPK(30),AK2(30),AKF(30),AKFU(30),GAXR,GAXZ
    COMMON /AC/ EN(100,5),ET(5),AA1(5),AA2(5),AA3(5),AVE(5),NAC,NE,SAF
   1I,SAFV,SAFPSI,SAFRO,SAFAX
    COMMON /CPL/ NI,NJ,NK,EP,ZLE,GAM,SM,N1,N2,N3,N4,N5,NVAC,HS,HU,HV,P
   1I,RX,RY,E4,A1,A2,A3,A4,A5,A6,HU4,HV4,HUV,IC,IO,SA1,SA2,SA3,SE1,SE2
   2,SE3,DT,RA1,RA2,RA3,RE1,RE2,RE3,ENER,FAXIS,DG1,DG2,DG3,DH1,DH2,DH3
   3,NPLOT,I1
```

```
      COMMON /CVA/ NIV,FV1,FV2,NV,NP,OM,PM,HR,H1,H2,H3,C1,C2,N6,EVAC,ETO
     1T,A11,A22,A12,HR4,DTB,FAXV,OM1,OM2,PM1,PM2
      A22=0.0
      AA1(4)=0.0
      DSU=1.0/(2.0*HR*HU)
      AT=0.01
      IF (IT.GT.0) AT=0.1
      IT=1
C     COMPUTE COEFFICIENTS FOR VACUUM EQUATIONS
      DO 30 J=2,N1
      X1=0.5*(RR(J)+RR(J-1)-RB(J)-RB(J-1))
      X2=0.5*(ZZ(J)+ZZ(J-1)-ZB(J)-ZB(J-1))
      X3=0.5*(RBU(J)+RBU(J-1))
      X4=0.5*(ZBU(J)+ZBU(J-1))
      X5=0.5*(RU(J)+RU(J-1))
      X6=0.5*(ZU(J)+ZU(J-1))
      X7=RR(J)-RB(J)
      X8=ZZ(J)-ZB(J)
      DO 10 I=1,N6
      S=(I-0.5)*HR
      S1=1.0-S
      Y1=S1*X3+S*X5
      Y2=S1*X4+S*X6
      AK=1.0+EP*(0.5*(RB(J)+RB(J-1))+S*X1)
      E(I,J)=AK*(Y1*Y1+Y2*Y2)/(X1*Y2-X2*Y1)
      Y1=S1*RBU(J)+S*RU(J)
      Y2=S1*ZBU(J)+S*ZU(J)
      AK=1.0+EP*(RB(J)+S*X7)
   10 G(I,J)=-AK*(Y1*X7+Y2*X8)/(X7*Y2-X8*Y1)
      DO 20 I=1,NIV
      S=(I-1)*HR
      S1=1.0-S
      Y1=S1*RBU(J)+S*RU(J)
      Y2=S1*ZBU(J)+S*ZU(J)
      AK=1.0+EP*(RB(J)+S*X7)
      F(I,J)=AK*(X7*X7+X8*X8)/(X7*Y2-X8*Y1)
      AK=1.0+EP*(0.5*(RB(J)+RB(J-1))+S*X1)
   20 D(I,J)=(X1*(S1*X4+S*X6)-X2*(S1*X3+S*X5))/AK
      F(1,J)=0.5*F(1,J)
      F(NIV,J)=0.5*F(NIV,J)
      D(1,J)=0.5*D(1,J)
      D(NIV,J)=0.5*D(NIV,J)
   30 CONTINUE
      DO 40 J=2,N1
      DO 40 I=1,NIV
   40 A22=A22+D(I,J)
      A22=(A22*HR*HU)/ZLE
      DO 50 I=1,N6
      E(I,1)=E(I,N1)
      F(I,1)=F(I,N1)
   50 G(I,1)=G(I,N1)
      F(NIV,1)=F(NIV,N1)
   60 NCO=NCO+1
      ERVA=0.0
      AA3(4)=AA1(4)
      AA1(4)=0.0
C     COMPUTE SOR ITERATION
```

```
      DO 80 J=2,N1
      SUM1=H1*E(1,J)*PA(2,J)+H2*(F(1,J)*PA(1,J+1)+F(1,J-1)*PA(1,J-1))+DS
     1U*(G(1,J)*PA(2,J+1)-G(1,J-1)*PA(2,J-1))
      SUM2=H1*E(1,J)+H2*(F(1,J)+F(1,J-1))+DSU*(G(1,J)-G(1,J-1))
      X1=SUM1-PA(1,J)*SUM2
      AA1(4)=AA1(4)+X1*X1
      PAS=PM1*PA(1,J)+(OM1*SUM1)/SUM2
      AA=ABS(X1)
      ERVA=AMAX1(ERVA,AA)
      PA(1,J)=PAS
      DO 70 I=2,N6
      SUM1=H1*(E(I,J)*PA(I+1,J)+E(I-1,J)*PA(I-1,J))+H2*(F(I,J)*PA(I,J+1)
     1+F(I,J-1)*PA(I,J-1))+DSU*(G(I,J)*PA(I+1,J+1)-G(I,J-1)*PA(I+1,J-1)-
     2G(I-1,J)*PA(I-1,J+1)+G(I-1,J-1)*PA(I-1,J-1))
      SUM2=H1*(E(I,J)+E(I-1,J))+H2*(F(I,J)+F(I,J-1))+DSU*(G(I,J)-G(I,J-1
     1)-G(I-1,J)+G(I-1,J-1))
      X1=SUM1-PA(I,J)*SUM2
      AA1(4)=AA1(4)+X1*X1
      PAS=PM1*PA(I,J)+(OM1*SUM1)/SUM2
      AA=ABS(X1)
      ERVA=AMAX1(ERVA,AA)
   70 PA(I,J)=PAS
      SUM1=H1*E(N6,J)*PA(N6,J)+H2*(F(NIV,J)*PA(NIV,J+1)+F(NIV,J-1)*PA(NI
     1V,J-1))+DSU*(-G(N6,J)*PA(N6,J+1)+G(N6,J-1)*PA(N6,J-1))
      SUM2=H1*E(N6,J)+H2*(F(NIV,J)+F(NIV,J-1))+DSU*(-G(N6,J)+G(N6,J-1))
      X1=SUM1-PA(NIV,J)*SUM2
      AA1(4)=AA1(4)+X1*X1
      PAS=PM1*PA(NIV,J)+(OM1*SUM1)/SUM2
      AA=ABS(X1)
      ERVA=AMAX1(ERVA,AA)
   80 PA(NIV,J)=PAS
      DO 90 I=1,NIV
      PA(I,1)=PA(I,N1)-1.0
   90 PA(I,N2)=PA(I,2)+1.0
      IF (NCO.GT.300) GO TO 130
      IF (ERVA.GT.AT) GO TO 60
      IF (NCO.LT.NV) GO TO 60
C     COMPUTE INDUCTANCE MATRIX
      A11=0.0
      DO 120 J=2,N1
      DO 100 I=1,N6
      X1=PA(I+1,J)-PA(I,J)
      X2=PA(I+1,J+1)-PA(I,J)
      X3=PA(I,J+1)-PA(I+1,J)
  100 A11=A11+H1*E(I,J)*X1*X1+DSU*G(I,J)*(X2*X2-X3*X3)
      DO 110 I=1,NIV
      X1=PA(I,J+1)-PA(I,J)
  110 A11=A11+H2*F(I,J)*X1*X1
  120 CONTINUE
      A11=ZLE*HR*HU*A11
      C1=FV1/A11
      C2=FV2/A22
      EVNE=0.5*(C1*FV1+C2*FV2)
      GO TO 140
  130 PRINT 150
      CALL ASOUT
      STOP
```

```
 140 CONTINUE
     AA1(4)=AA1(4)*HU*HR
     AA3(4)=AA3(4)*HU*HR
     AA2(4)=0.5*AA1(4)
     RETURN
 150 FORMAT (////,6X34HLAPLACE EQUATION DOES NOT CONVERGE)
     END



     SUBROUTINE ASBO (ERBO,ERBO1)
C    COMPUTES ONE ITERATION OF FREE BOUNDARY EQUATION
     COMMON /AUX/ R2,Z2,RA,ZA,X(30),R(10,30),AL(10,30),PA(10,30),D(10,3
    10),E(10,30),F(10,30),G(10,30),R1(10,30),Z1(10,30),GR(10,30),GZ(10,
    230),DD(10,30),QR(10,30),QZ(10,30),A(30),B(30),C(30),H(30),RR(30),Z
    3Z(30),RU(30),ZU(30),RB(30),ZB(30),RBU(30),ZBU(30),XA(30),XC(30),XP
    4D(30),XPK(30),AK2(30),AKF(30),AKFU(30),GAXR,GAXZ
     COMMON /INP/ Q(10),QT(10),AM(10),PR(10),QQ(10),PC(10),TC(10),BPP(1
    10),BTP(10),BET(10)
     COMMON /CPL/ NI,NJ,NK,EP,ZLE,GAM,SM,N1,N2,N3,N4,N5,NVAC,HS,HU,HV,P
    1I,RX,RY,E4,A1,A2,A3,A4,A5,A6,HU4,HV4,HUV,IC,IO,SA1,SA2,SA3,SE1,SE2
    2,SE3,DT,RA1,RA2,RA3,RE1,RE2,RE3,ENER,FAXIS,DG1,DG2,DG3,DH1,DH2,DH3
    3,NPLOT,I1
     COMMON /CVA/ NIV,FV1,FV2,NV,NP,OM,PM,HR,H1,H2,H3,C1,C2,N6,EVAC,ETO
    1T,A11,A22,A12,HR4,DTB,FAXV,OM1,OM2,PM1,PM2
     C11=C1*C1
     C22=C2*C2
     Z22=ZLE*ZLE
     ERBO=0.0
     SUM1=0.0
     SUM2=0.0
     HS8=8.0*HS
     PB=0.5*(3.0*PR(N3)-PR(N3-1))
     Y1=3.0-4.0*R(N3,2)*R(N3,2)+R(N3-1,2)*R(N3-1,2)
     Q2=Q(NI)*Q(NI)
C    COMPUTE COEFFICIENTS FOR LAX-WENDROFF METHOD
     DO 10 J=2,N1
     GU2=(PA(1,J+1)-PA(1,J))*(PA(1,J+1)-PA(1,J))*H2
     AU=RBU(J)
     BU=ZBU(J)
     AK=1.0+EP*RB(J)
     ABU=AU*AU+BU*BU
     Y2=3.0-4.0*R(N3,J+1)*R(N3,J+1)+R(N3-1,J+1)*R(N3-1,J+1)
     PL=H(J)*(Y1+Y2)/HS8
     Y1=Y2
     X1=RR(J)-R2
     X2=ZZ(J)-Z2
     PU2=(AL(NI,J+1)-AL(NI,J))*(AL(NI,J+1)-AL(NI,J))*H2
     PAS1=C11*GU2/ABU
     BV2=PAS1+C22/(Z22*AK*AK)
     BP2=(AK*AK*PU2+Q2*ABU)/(AK*AK*PL*PL)
     AK2(J)=AK*(0.5*(BP2-BV2)+PB)
     X3=RU(J)
     X4=ZU(J)
     PAS2=AU*X3+BU*X4
```

```
      PAS3=Q2/(PL*PL*AK*AK)
      PAS4=(X(J)+X(J+1))*(X1*X4-X2*X3)+(R2-RA)*X4-(Z2-ZA)*X3
      R30=RB(J)-RA
      Z30=ZB(J)-ZA.
      PAS5=BP2/(R30*BU-Z30*AU)
      PAS6=PAS3+PAS1/ABU
      AKF(J)=EP*X1*(AK2(J)/AK-PAS3*ABU+C22/(Z22*AK*AK))+AK*(PAS6*PAS2-PA
     1S5*PAS4)
   10 AKFU(J)=AK*(PAS6*(X1*AU+X2*BU)-PAS5*((R2-RA)*X2-(Z2-ZA)*X1))
      AK2(1)=AK2(N1)
      AKF(1)=AKF(N1)
      AKFU(1)=AKFU(N1)
C     COMPUTE LAX-WENDROFF ITERATION
      DO 20 J=2,N1
      X1=0.5*(AK2(J)+AK2(J-1))
      X2=0.5*(AKF(J)+AKF(J-1))
      X3=0.5*(AKFU(J)+AKFU(J-1))
      X4=(AK2(J)-AK2(J-1))/HU
      X5=X1+0.5*DTB*(X1*X2+X3*X4)
      SUM1=SUM1+0.5*X5*(ZBU(J)+ZBU(J-1))*(1.0-X(J))
      SUM2=SUM2-0.5*X5*(RBU(J)+RBU(J-1))*(1.0-X(J))
      AA=ABS(X5)
      ERBO=AMAX1(ERBO,AA)
      X(J)=X(J)+DTB*X5
   20 CONTINUE
      SUM1=SUM1*HU
      SUM2=SUM2*HU
      X(1)=X(N1)
      X(N2)=X(2)
C     COMPUTE VACUUM AXIS ITERATION
      R2=R2+DTB*SUM1*FAXV
      Z2=Z2+DTB*SUM2*FAXV
      AA=ABS(SUM1)
      BB=ABS(SUM2)
      ERBO1=AMAX1(AA,BB)
      RETURN
      END



      SUBROUTINE TSOR (ERVA,EVNE,IT)
C     COMPUTES NV ITERATIONS OF 3-D VACUUM EQUATIONS
      COMMON RO(10,30,30),AL(10,30,30),XO(10,30,30),XL(10,30,30),R(30,30
     1),Z(30,30),RU(30,30),ZU(30,30),RV(30,30),ZV(30,30),X(30,30),RA(30)
     2,ZA(30),RN(30),ZN(30),RB1(30),RB2(30),ZB1(30),ZB2(30),RBU1(30),RBU
     32(30),ZBU1(30),ZBU2(30),RBV1(30),RBV2(30),ZBV1(30),ZBV2(30),HB1(30
     4),HB2(30)
      COMMON /AUX/ VA(10,30),VB1(10,30),VB2(10,30),VC(10,30),VD(10,30),V
     1E1(10,30),VE2(10,30),VL1(10,30),VL2(10,30)
      COMMON /POT/ RVA(30),ZVA(30),BPV(10),BTV(10),PT(10,30,30),PP(10,30
     1,30)
      COMMON /AC/ EN(100,5),ET(5),AA1(5),AA2(5),AA3(5),AVE(5),NAC,NE,SAF
     1I,SAFV,SAFPSI,SAFRO,SAFAX
      COMMON /CPL/ NI,NJ,NK,EP,ZLE,GAM,SM,N1,N2,N3,N4,N5,NVAC,HS,HU,HV,P
     1I,RX,RY,E4,A1,A2,A3,A4,A5,A6,HU4,HV4,HUV,IC,IO,SA1,SA2,SA3,SE1,SE2
```

```
      2,SE3,DT,RA1,RA2,RA3,RE1,RE2,RE3,ENER,FAXIS,DG1,DG2,DG3,DH1,DH2,DH3
      3,NPLOT,I1
      COMMON /CVA/ NIV,FV1,FV2,NV,NP,OM,PM,HR,H1,H2,H3,C1,C2,N6,EVAC,ETO
      1T,A11,A22,A12,HR4,DT8,FAXV,OM1,OM2,PM1,PM2
      NCO=0
      AT=0.01
      IF (IT.GT.0) AT=0.1
      IT=1
      IF (NPLOT.LT.0) GO TO 30
      DO 10 I=1,NIV
      BPV(I)=0.0
   10 BTV(I)=0.0
      DO 20 I=1,NIV
      DO 20 J=1,N2
      DO 20 K=1,N5
   20 PP(I,J,K)=C1*PP(I,J,K)+C2*PT(I,J,K)
   30 CONTINUE
   40 CONTINUE
      NCO=NCO+1
      A11=0.0
      A22=0.0
      A12=0.0
      AA3(4)=AA1(4)
      AA3(5)=AA1(5)
      AA1(4)=0.0
      AA1(5)=0.0
      ERVA=0.0
      ZZ=ZLE*ZLE
      DU=1.0/(4.0*HU)
      DV=1.0/(4.0*HV)
      DSU=1.0/(2.0*HR*HU)
      DSV=1.0/(2.0*HR*HV)
      DVU=1.0/(2.0*HV*HU)
      DUV=DVU
      H4=1.0/(HV*HV)
      CALL CBO (1)
      CALL CV1 (1)
C     COEFFICIENT MATRICES ARE DENOTED BY 1 WHEN EVALUATED AT K+1/2
C     AND BY 2 WHEN EVALUATED AT K-1/2
      DO 160 K=2,N4
      DO 60 J=1,N1
      RB2(J)=RB1(J)
      ZB2(J)=ZB1(J)
      RBU2(J)=RBU1(J)
      ZBU2(J)=ZBU1(J)
      RBV2(J)=RBV1(J)
      ZBV2(J)=ZBV1(J)
      DO 50 I=1,NIV
      VB2(I,J)=VB1(I,J)
   50 VL2(I,J)=VL1(I,J)
      DO 60 I=1,N6
   60 VE2(I,J)=VE1(I,J)
C     EVALUATES BOUNDARY QUANTITIES AT K+1/2
      CALL CBO (K)
C     EVALUATES VACUUM EQUATIONS COEFFICIENTS AT K+1/2
      CALL CV1 (K)
C     EVALUATES VACUUM EQUATIONS COEFFICIENTS AT K
```

```
        CALL CV (K)
        IF (NPLOT.LT.0) GO TO 80
C       COMPUTE FINAL VALUES OF VACUUM MAGNETIC FIELD
        DO 70 J=2,N1
        DO 70 I=1,N6
        S=(I-0.5)*HR
        S1=1.0-S
        X1=S1*RBU1(J)+S*RU(J,K)
        X2=S1*RBV1(J)+S*RV(J,K)
        X3=S1*ZBU1(J)+S*ZU(J,K)
        X4=S1*ZBV1(J)+S*ZV(J,K)
        AK=1.0+EP*(RB1(J)+S*(R(J,K)-RB1(J)))
        PU=JU*(PP(I,J+1,K)+PP(I,J+1,K+1)+PP(I+1,J+1,K)+PP(I+1,J+1,K+1)-PP(
       1I,J,K)-PP(I,J,K+1)-PP(I+1,J,K)-PP(I+1,J,K+1))
        PV=DV*(PP(I,J,K+1)+PP(I+1,J,K+1)+PP(I,J+1,K+1)+PP(I+1,J+1,K+1)-PP(
       1I,J,K)-PP(I+1,J,K)-PP(I,J+1,K)-PP(I+1,J+1,K))
        X1=SQRT(X1*X1+X3*X3)
        X2=SQRT(X2*X2+X4*X4+ZZ*AK*AK)
C       POLOIDAL AND TOROIDAL VACUUM FIELDS
        PP(I,J,K)=PU/X1
        PT(I,J,K)=PV/X2
        BPV(I)=BPV(I)+PP(I,J,K)
   70   BTV(I)=BTV(I)+PT(I,J,K)
        GO TO 160
   80 CONTINUE
C       SOR ITERATION
C       ITERATION FOR FREE BOUNDARY POINTS
        DO 90 J=2,N1
        SUM1=VA(1,J)*PP(2,J,K)*H1+H4*(VB1(1,J)*PP(1,J,K+1)+VB2(1,J)*PP(1,J
       1,K-1))+H2*(VC(1,J)*PP(1,J+1,K)+VC(1,J-1)*PP(1,J-1,K))+DSU*(VD(1,J)
       2*PP(2,J+1,K)-VD(1,J-1)*PP(2,J-1,K))+DSV*(VE1(1,J)*PP(2,J,K+1)-VE2(
       31,J)*PP(2,J,K-1))+DUV*(VL1(1,J)*PP(1,J+1,K+1)-VL1(1,J-1)*PP(1,J-1,
       4K+1)-VL2(1,J)*PP(1,J+1,K-1)+VL2(1,J-1)*PP(1,J-1,K-1))
        SUM2=VA(1,J)*PT(2,J,K)*H1+H4*(VB1(1,J)*PT(1,J,K+1)+VB2(1,J)*PT(1,J
       1,K-1))+H2*(VC(1,J)*PT(1,J+1,K)+VC(1,J-1)*PT(1,J-1,K))+DSU*(VD(1,J)
       2*PT(2,J+1,K)-VD(1,J-1)*PT(2,J-1,K))+DSV*(VE1(1,J)*PT(2,J,K+1)-VE2(
       31,J)*PT(2,J,K-1))+DUV*(VL1(1,J)*PT(1,J+1,K+1)-VL1(1,J-1)*PT(1,J-1,
       4K+1)-VL2(1,J)*PT(1,J+1,K-1)+VL2(1,J-1)*PT(1,J-1,K-1))
        SUM3=H1*VA(1,J)+H4*(VB1(1,J)+VB2(1,J))+H2*(VC(1,J)+VC(1,J-1))+DSU*
       1(VD(1,J)-VD(1,J-1))+DSV*(VE1(1,J)-VE2(1,J))+DUV*(VL1(1,J)-VL1(1,J-
       21)-VL2(1,J)+VL2(1,J-1))
        PAS1=PM1*PP(1,J,K)+(OM1*SUM1)/SUM3
        PAS2=PM2*PT(1,J,K)+(OM2*SUM2)/SUM3
        AA=ABS(PAS1-PP(1,J,K))/DT
        BB=ABS(PAS2-PT(1,J,K))/DT
        AA1(4)=AA1(4)+AA*AA
        AA1(5)=AA1(5)+BB*BB
        ERVA=AMAX1(ERVA,AA,BB)
        PP(1,J,K)=PAS1
        PT(1,J,K)=PAS2
   90 CONTINUE
C       PERIODICITY CONDITIONS
        PP(1,1,K)=PP(1,N1,K)-1.0
        PP(1,N2,K)=PP(1,2,K)+1.0
        PT(1,1,K)=PT(1,N1,K)
        PT(1,N2,K)=PT(1,2,K)
C       ITERATION FOR INTERIOR POINTS
```

```
      DO 110 I=2,N6
      DO 100 J=2,N1
      SUM1=H1*(VA(I,J)*PP(I+1,J,K)+VA(I-1,J)*PP(I-1,J,K))+H4*(VB1(I,J)*P
     1P(I,J,K+1)+VB2(I,J)*PP(I,J,K-1))+H2*(VC(I,J)*PP(I,J+1,K)+VC(I,J-1)
     2*PP(I,J-1,K))+DSU*(VD(I,J)*PP(I+1,J+1,K)-VD(I,J-1)*PP(I+1,J-1,K)-V
     3D(I-1,J)*PP(I-1,J+1,K)+VD(I-1,J-1)*PP(I-1,J-1,K))+DSV*(VE1(I,J)*PP
     4(I+1,J,K+1)-VE1(I-1,J)*PP(I-1,J,K+1)-VE2(I,J)*PP(I+1,J,K-1)+VE2(I-
     51,J)*PP(I-1,J,K-1))+DVU*(VL1(I,J)*PP(I,J+1,K+1)-VL1(I,J-1)*PP(I,J-
     61,K+1)-VL2(I,J)*PP(I,J+1,K-1)+VL2(I,J-1)*PP(I,J-1,K-1))
      SUM2=H1*(VA(I,J)*PT(I+1,J,K)+VA(I-1,J)*PT(I-1,J,K))+H4*(VB1(I,J)*P
     1T(I,J,K+1)+VB2(I,J)*PT(I,J,K-1))+H2*(VC(I,J)*PT(I,J+1,K)+VC(I,J-1)
     2*PT(I,J-1,K))+DSU*(VD(I,J)*PT(I+1,J+1,K)-VD(I,J-1)*PT(I+1,J-1,K)-V
     3D(I-1,J)*PT(I-1,J+1,K)+VD(I-1,J-1)*PT(I-1,J-1,K))+DSV*(VE1(I,J)*PT
     4(I+1,J,K+1)-VE1(I-1,J)*PT(I-1,J,K+1)-VE2(I,J)*PT(I+1,J,K-1)+VE2(I-
     51,J)*PT(I-1,J,K-1))+DVU*(VL1(I,J)*PT(I,J+1,K+1)-VL1(I,J-1)*PT(I,J-
     61,K+1)-VL2(I,J)*PT(I,J+1,K-1)+VL2(I,J-1)*PT(I,J-1,K-1))
      SUM3=H1*(VA(I,J)+VA(I-1,J))+H4*(VB1(I,J)+VB2(I,J))+H2*(VC(I,J)+VC(
     1I,J-1))+DSU*(VD(I,J)-VD(I,J-1)-VD(I-1,J)+VD(I-1,J-1))+DSV*(VE1(I,J
     2)-VE1(I-1,J)-VE2(I,J)+VE2(I-1,J))+DVU*(VL1(I,J)-VL1(I,J-1)-VL2(I,J
     3)+VL2(I,J-1))
      PAS1=PM1*PP(I,J,K)+(OM1*SUM1)/SUM3
      PAS2=PM2*PT(I,J,K)+(OM2*SUM2)/SUM3
      AA=ABS(PAS1-PP(I,J,K))/DT
      BB=ABS(PAS2-PT(I,J,K))/DT
      AA1(4)=AA1(4)+AA*AA
      AA1(5)=AA1(5)+BB*BB
      ERVA=AMAX1(ERVA,AA,BB)
      PP(I,J,K)=PAS1
      PT(I,J,K)=PAS2
  100 CONTINUE
C     PERIODICITY CONDITIONS
      PP(I,1,K)=PP(I,N1,K)-1.0
      PP(I,N2,K)=PP(I,2,K)+1.0
      PT(I,1,K)=PT(I,N1,K)
  110 PT(I,N2,K)=PT(I,2,K)
C     ITERATION FOR OUTER WALL POINTS
      DO 120 J=2,N1
      SUM1=H1*VA(N6,J)*PP(N6,J,K)+H4*(VB1(NIV,J)*PP(NIV,J,K+1)+VB2(NIV,J
     1)*PP(NIV,J,K-1))+H2*(VC(NIV,J)*PP(NIV,J+1,K)+VC(NIV,J-1)*PP(NIV,J-
     21,K))+DSU*(-VD(N6,J)*PP(N6,J+1,K)+VD(N6,J-1)*PP(N6,J-1,K))+DSV*(-V
     3E1(N6,J)*PP(N6,J,K+1)+VE2(N6,J)*PP(N6,J,K-1))+DVU*(VL1(NIV,J)*PP(N
     4IV,J+1,K+1)-VL1(NIV,J-1)*PP(NIV,J-1,K+1)-VL2(NIV,J)*PP(NIV,J+1,K-1
     5)+VL2(NIV,J-1)*PP(NIV,J-1,K-1))
      SUM2=H1*VA(N6,J)*PT(N6,J,K)+H4*(VB1(NIV,J)*PT(NIV,J,K+1)+VB2(NIV,J
     1)*PT(NIV,J,K-1))+H2*(VC(NIV,J)*PT(NIV,J+1,K)+VC(NIV,J-1)*PT(NIV,J-
     21,K))+DSU*(-VD(N6,J)*PT(N6,J+1,K)+VD(N6,J-1)*PT(N6,J-1,K))+DSV*(-V
     3E1(N6,J)*PT(N6,J,K+1)+VE2(N6,J)*PT(N6,J,K-1))+DVU*(VL1(NIV,J)*PT(N
     4IV,J+1,K+1)-VL1(NIV,J-1)*PT(NIV,J-1,K+1)-VL2(NIV,J)*PT(NIV,J+1,K-1
     5)+VL2(NIV,J-1)*PT(NIV,J-1,K-1))
      SUM3=H1*VA(N6,J)+H4*(VB1(NIV,J)+VB2(NIV,J))+H2*(VC(NIV,J)+VC(NIV,J
     1-1))+DSU*(-VD(N6,J)+VD(N6,J-1))+DSV*(-VE1(N6,J)+VE2(N6,J))+DVU*(VL
     21(NIV,J)-VL1(NIV,J-1)-VL2(NIV,J)+VL2(NIV,J-1))
      PAS1=PM1*PP(NIV,J,K)+(OM1*SUM1)/SUM3
      PAS2=PM2*PT(NIV,J,K)+(OM2*SUM2)/SUM3
      AA=ABS(PAS1-PP(NIV,J,K))/DT
      BB=ABS(PAS2-PT(NIV,J,K))/DT
      AA1(4)=AA1(4)+AA*AA
```

```
          AA1(5)=AA1(5)+BB*BB
          ERVA=AMAX1(ERVA,AA,BB)
          PP(NIV,J,K)=PAS1
          PT(NIV,J,K)=PAS2
      120 CONTINUE
C         PERIODICITY CONDITIONS
          PP(NIV,1,K)=PP(NIV,N1,K)-1.0
          PP(NIV,N2,K)=PP(NIV,2,K)+1.0
          PT(NIV,1,K)=PT(NIV,N1,K)
          PT(NIV,N2,K)=PT(NIV,2,K)
          IF (NCO.LT.NV) GO TO 160
C         COMPUTE INDUCTANCE MATRIX
          DO 150 J=2,N1
          DO 130 I=1,N6
          Y1=PP(I+1,J,K)-PP(I,J,K)
          T1=PT(I+1,J,K)-PT(I,J,K)
          Y4=PP(I+1,J+1,K)-PP(I,J,K)
          T4=PT(I+1,J+1,K)-PT(I,J,K)
          Y5=PP(I+1,J,K)-PP(I,J+1,K)
          T5=PT(I+1,J,K)-PT(I,J+1,K)
          Y6=PP(I+1,J,K)-PP(I,J,K-1)
          T6=PT(I+1,J,K)-PT(I,J,K-1)
          Y7=PP(I+1,J,K-1)-PP(I,J,K)
          T7=PT(I+1,J,K-1)-PT(I,J,K)
          X1=41*VA(I,J)
          X2=DSU*VD(I,J)
          X3=DSV*VE2(I,J)
          A11=A11+X1*Y1*Y1+X2*(Y4*Y4-Y5*Y5)+X3*(Y6*Y6-Y7*Y7)
          A22=A22+X1*T1*T1+X2*(T4*T4-T5*T5)+X3*(T6*T6-T7*T7)
      130 A12=A12+X1*Y1*T1+X2*(Y4*T4-Y5*T5)+X3*(Y6*T6-Y7*T7)
          DO 140 I=1,NIV
          Y2=PP(I,J,K)-PP(I,J,K-1)
          T2=PT(I,J,K)-PT(I,J,K-1)
          Y3=PP(I,J+1,K)-PP(I,J,K)
          T3=PT(I,J+1,K)-PT(I,J,K)
          Y8=PP(I,J+1,K)-PP(I,J,K-1)
          T8=PT(I,J+1,K)-PT(I,J,K-1)
          Y9=PP(I,J,K)-PP(I,J+1,K-1)
          T9=PT(I,J,K)-PT(I,J+1,K-1)
          X1=H4*VB2(I,J)
          X2=H2*VC(I,J)
          X3=DVU*VL2(I,J)
          A11=A11+X1*Y2*Y2+X2*Y3*Y3+X3*(Y8*Y8-Y9*Y9)
          A22=A22+X1*T2*T2+X2*T3*T3+X3*(T8*T8-T9*T9)
      140 A12=A12+X1*Y2*T2+X2*Y3*T3+X3*(Y8*T6-Y9*T9)
      150 CONTINUE
      160 CONTINUE
          IF (NPLOT.LT.0) GO TO 180
C         AVERAGE POLOIDAL AND TOROIDAL VACUUM FIELDS
          DO 170 I=1,N6
          BPV(I)=BPV(I)*HU*HV
      170 BTV(I)=BTV(I)*HU*HV
          GO TO 210
      180 CONTINUE
C         PERIODICITY CONDITIONS
          DO 190 I=1,NIV
          DO 190 J=1,N2
```

```
      PP(I,J,1)=PP(I,J,N4)
      PP(I,J,N5)=PP(I,J,2)
      PT(I,J,1)=PT(I,J,N4)-1.0
  190 PT(I,J,N5)=PT(I,J,2)+1.0
      IF (NCO.GT.300) GO TO 200
      IF (ERVA.GT.AT) GO TO 40
      IF (NCO.LT.NV) GO TO 40
      HURV=HU*HR*HV
      A11=A11*HURV
      A12=A12*HURV
      A22=A22*HURV
C     FUNCTIONAL FOR COMPUTATION OF OMEGA
      AA1(4)=AA1(4)*HURV
      AA1(5)=AA1(5)*HURV
      AA3(4)=AA3(4)*HURV
      AA3(5)=AA3(5)*HURV
      AA2(4)=0.5*AA1(4)
      AA2(5)=0.5*AA1(5)
      DEL=A11*A22-A12*A12
C     CURRENTS AND VACUUM ENERGY
      C1=(FV1*A22-FV2*A12)/DEL
      C2=(FV2*A11-FV1*A12)/DEL
      EVNE=0.5*(C1*FV1+C2*FV2)
      GO TO 210
  200 PRINT 220
      STOP
  210 CONTINUE
      RETURN
  220 FORMAT (////,6X34HLAPLACE EQUATION DOES NOT CONVERGE)
      END



      SUBROUTINE CV1 (K)
C     EVALUATES VACUUM EQUATIONS COEFFICIENTS AT K+1/2
      COMMON RO(10,30,30),AL(10,30,30),XO(10,30,30),XL(10,30,30),R(30,30
     1),Z(30,30),RU(30,30),ZU(30,30),RV(30,30),ZV(30,30),X(30,30),RA(30)
     2,ZA(30),RN(30),ZN(30),RB1(30),RB2(30),ZB1(30),ZB2(30),RBU1(30),RBU
     32(30),ZBU1(30),ZBU2(30),RBV1(30),RBV2(30),ZBV1(30),ZBV2(30),HB1(30
     4),HB2(30)
      COMMON /AUX/ VA(10,30),VB1(10,30),VB2(10,30),VC(10,30),VD(10,30),V
     1E1(10,30),VE2(10,30),VL1(10,30),VL2(10,30)
      COMMON /CPL/ NI,NJ,NK,EP,ZLE,GAM,SM,N1,N2,N3,N4,N5,NVAC,HS,HU,HV,P
     1I,RX,RY,E4,A1,A2,A3,A4,A5,A6,HU4,HV4,HUV,IC,IO,SA1,SA2,SA3,SE1,SE2
     2,SE3,DT,RA1,RA2,RA3,RE1,RE2,RE3,ENER,FAXIS,DG1,DG2,DG3,DH1,DH2,DH3
     3,NPLOT,I1
      COMMON /CVA/ NIV,FV1,FV2,NV,NP,OM,PM,HR,H1,H2,H3,C1,C2,N6,EVAC,ETO
     1T,A11,A22,A12,HR4,DTB,FAXV,OM1,OM2,PM1,PM2
      EP2=0.5*EP
      DO 30 J=2,N1
      X1=0.5*(R(J,K)+R(J-1,K)-RB1(J)-RB1(J-1))
      X2=0.5*(Z(J,K)+Z(J-1,K)-ZB1(J)-ZB1(J-1))
      X3=0.5*(RBU1(J)+RBU1(J-1))
      X4=0.5*(ZBU1(J)+ZBU1(J-1))
      X5=0.5*(RU(J,K)+RU(J-1,K))
```

```
      X6=0.5*(ZU(J,K)+ZU(J-1,K))
      X7=0.5*(RBV1(J)+RBV1(J-1))
      X8=0.5*(ZBV1(J)+ZBV1(J-1))
      X9=0.5*(RV(J,K)+RV(J-1,K))
      X10=0.5*(ZV(J,K)+ZV(J-1,K))
      DO 10 I=1,N6
      S=(I-0.5)*HR
      S1=1.0-S
      AKL=ZLE*(1.0+EP2*(S1*(RB1(J)+RB1(J-1))+S*(R(J,K)+R(J-1,K))))
   10 VE1(I,J)=((S1*X3+S*X5)*(S1*X8+S*X10)-(S1*X7+S*X9)*(S1*X4+S*X6))/AK
     1L
      DO 20 I=1,NIV
      S=(I-1)*HR
      S1=1.0-S
      AKL=ZLE*(1.0+EP2*(S1*(RB1(J)+RB1(J-1))+S*(R(J,K)+R(J-1,K))))
      VB1(I,J)=(X1*(S1*X4+S*X6)-X2*(S1*X3+S*X5))/AKL
      AKL=ZLE*(1.0+EP*(S1*RB1(J)+S*R(J,K)))
   20 VL1(I,J)=((S1*RBV1(J)+S*RV(J,K))*(Z(J,K)-ZB1(J))-(S1*ZBV1(J)+S*ZV(
     1J,K))*(R(J,K)-RB1(J)))/AKL
      VB1(1,J)=0.5*VB1(1,J)
      VB1(NIV,J)=0.5*VB1(NIV,J)
      VL1(1,J)=0.5*VL1(1,J)
      VL1(NIV,J)=0.5*VL1(NIV,J)
   30 CONTINUE
      DO 40 I=1,NIV
      VB1(I,1)=VB1(I,N1)
   40 VL1(I,1)=VL1(I,N1)
      DO 50 I=1,N6
   50 VE1(I,1)=VE1(I,N1)
      RETURN
      END



      SUBROUTINE CV (K)
C     EVALUATES VACUUM EQUATIONS COEFFICIENTS AT K
      COMMON RO(10,30,30),AL(10,30,30),XO(10,30,30),XL(10,30,30),R(30,30
     1),Z(30,30),RU(30,30),ZU(30,30),RV(30,30),ZV(30,30),X(30,30),RA(30)
     2,ZA(30),RN(30),ZN(30),RB1(30),RB2(30),ZB1(30),ZB2(30),RBU1(30),RBU
     32(30),ZBU1(30),ZBU2(30),RBV1(30),RBV2(30),ZBV1(30),ZBV2(30),HB1(30
     4),HB2(30)
      COMMON /CPL/ NI,NJ,NK,EP,ZLE,GAM,SM,N1,N2,N3,N4,N5,NVAC,HS,HU,HV,P
     1I,RX,RY,E4,A1,A2,A3,A4,A5,A6,HU4,HV4,HUV,IC,IO,SA1,SA2,SA3,SE1,SE2
     2,SE3,DT,RA1,RA2,RA3,RE1,RE2,RE3,ENER,FAXIS,DG1,DG2,DG3,DH1,DH2,DH3
     3,NPLOT,I1
      COMMON /CVA/ NIV,FV1,FV2,NV,NP,OM,PM,HR,H1,H2,H3,C1,C2,N6,EVAC,ETO
     1T,A11,A22,A12,HR4,DTB,FAXV,OM1,OM2,PM1,PM2
      COMMON /AUX/ VA(10,30),VB1(10,30),VB2(10,30),VC(10,30),VD(10,30),V
     1E1(10,30),VE2(10,30),VL1(10,30),VL2(10,30)
      Z1=0.5*(R(1,K)+R(1,K-1)-RB1(1)-RB2(1))
      Z2=0.5*(Z(1,K)+Z(1,K-1)-ZB1(1)-ZB2(1))
      Z3=0.5*(RBU1(1)+RBU2(1))
      Z4=0.5*(ZBU1(1)+ZBU2(1))
      Z5=0.5*(RU(1,K)+RU(1,K-1))
      Z6=0.5*(ZU(1,K)+ZU(1,K-1))
```

```
      Z7=0.5*(RBV1(1)+RBV2(1))
      Z8=0.5*(ZBV1(1)+ZBV2(1))
      Z9=0.5*(RV(1,K)+RV(1,K-1))
      Z10=0.5*(ZV(1,K)+ZV(1,K-1))
      DO 30 J=2,N1
      X1=0.5*(R(J,K)+R(J,K-1)-RB1(J)-RB2(J))
      X2=0.5*(Z(J,K)+Z(J,K-1)-ZB1(J)-ZB2(J))
      X3=0.5*(RBU1(J)+RBU2(J))
      X4=0.5*(ZBU1(J)+ZBU2(J))
      X5=0.5*(RU(J,K)+RU(J,K-1))
      X6=0.5*(ZU(J,K)+ZU(J,K-1))
      X7=0.5*(RBV1(J)+RBV2(J))
      X8=0.5*(ZBV1(J)+ZBV2(J))
      X9=0.5*(RV(J,K)+RV(J,K-1))
      X10=0.5*(ZV(J,K)+ZV(J,K-1))
      DO 10 I=1,NIV
      S=(I-1)*HR
      S1=1.0-S
      Y1=(S1*X7+S*X9)*X2-(S1*X8+S*X10)*X1
      AKL=ZLE*(1.0+EP*(0.5*(RB1(J)+RB2(J))+S*X1))
      DEL=X1*(S1*X4+S*X6)-X2*(S1*X3+S*X5)
   10 VC(I,J)=(AKL*(X1*X1+X2*X2)+(Y1*Y1)/AKL)/DEL
      VC(1,J)=0.5*VC(1,J)
      VC(NIV,J)=0.5*VC(NIV,J)
      DO 20 I=1,N6
      S=(I-0.5)*HR
      S1=1.0-S
      Y1=S1*X3+S*X5
      Y2=S1*X4+S*X6
      Y3=S1*X7+S*X9
      Y4=S1*X8+S*X10
      AKL=ZLE*(1.0+EP*(0.5*(RB1(J)+RB2(J))+S*X1))
      VD(I,J)=((Y1*Y4-Y3*Y2)*(Y3*X2-Y4*X1)/AKL-(Y1*X1+Y2*X2)*AKL)/(X1*Y2
     1-X2*Y1)
      Y1=0.5*(S1*(X3+Z3)+S*(X5+Z5))
      Y2=0.5*(S1*(X4+Z4)+S*(X6+Z6))
      Y3=0.5*(S1*(X7+Z7)+S*(X9+Z9))
      Y4=0.5*(S1*(X8+Z8)+S*(X10+Z10))
      DEL=0.5*((X1+Z1)*Y2-(X2+Z2)*Y1)
      AKL=ZLE*(1.0+EP*(0.25*(RB1(J)+RB1(J-1)+RB2(J)+RB2(J-1))+S*0.5*(X1+
     1Z1)))
   20 VA(I,J)=(AKL*(Y1*Y1+Y2*Y2)+(Y1*Y4-Y2*Y3)*(Y1*Y4-Y2*Y3)/AKL)/DEL
      Z1=X1
      Z2=X2
      Z3=X3
      Z4=X4
      Z5=X5
      Z6=X6
      Z7=X7
      Z8=X8
      Z9=X9
   30 Z10=X10
      DO 40 I=1,N6
      VA(I,1)=VA(I,N1)
   40 VD(I,1)=VD(I,N1)
      DO 50 I=1,NIV
   50 VC(I,1)=VC(I,N1)
```

```
      RETURN
      END


      SUBROUTINE TBO (ERBO,ERBO1)
C     COMPUTES ONE ITERATICN OF 3-D FREE BOUNDARY EQUATION
      COMMON RO(10,30,30),AL(10,30,30),XO(10,30,30),XL(10,30,30),R(30,30
     1),Z(30,30),RU(30,30),ZU(30,30),RV(30,30),ZV(30,30),X(30,30),RA(30)
     2,ZA(30),RN(30),ZN(30),RB1(30),RB2(30),ZB1(30),ZB2(30),RBU1(30),RBU
     32(30),ZBU1(30),ZBU2(30),RBV1(30),RBV2(30),ZBV1(30),ZBV2(30),HB1(30
     4),HB2(30)
      COMMON /AUX/ AK(30,30),AKFU(30,30),AKFV(30,30),AKF(30,30)
      COMMON /POT/ RVA(30),ZVA(30),BPV(10),BTV(10),PT(10,30,30),PP(10,30
     1,30)
      COMMON /INP/ Q(10),QT(10),AM(10),PR(10),QQ(10),PC(10),TC(10),BPP(1
     10),BTP(10),BET(10)
      COMMON /CPL/ NI,NJ,NK,EP,ZLE,GAM,SM,N1,N2,N3,N4,N5,NVAC,HS,HU,HV,P
     11,RX,RY,E4,A1,A2,A3,A4,A5,A6,HU4,HV4,HUV,IC,IO,SA1,SA2,SA3,SE1,SE2
     2,SE3,DT,RA1,RA2,RA3,RE1,RE2,RE3,ENER,FAXIS,DG1,DG2,DG3,DH1,DH2,DH3
     3,NPLOT,I1
      COMMON /CVA/ NIV,FV1,FV2,NV,NP,OM,PM,HR,H1,H2,H3,C1,C2,N6,EVAC,ETO
     1T,A11,A22,A12,HR4,DTB,FAXV,OM1,OM2,PM1,PM2
      ERBO=0.0
      ERBO1=0.0
C     COMPUTE PRESSURE AND S DERIVATIVE OF R**2 AT THE BOUNDARY
      PB=0.5*(3.0*PR(N3)-PR(N3-1))
      DO 20 J=2,N1
      DO 10 K=2,N4
   10 AK(J,K)=3.0-4.0*RO(NI-1,J,K)*RO(NI-1,J,K)+RO(NI-2,J,K)*RO(NI-2,J,K
     1)
   20 AK(J,N5)=AK(J,2)
      DO 30 K=2,N5
   30 AK(N2,K)=AK(2,K)
      A7=1.0/(16.0*HS)
      ZZ=ZLE*ZLE
      C11=C1*C1
      C22=C2*C2
      C12=C1*C2
      C33=2.0*C12
      B1=1.0/(2.0*HU*HU)
      B2=1.0/(2.0*HV*HV)
      B3=1.0/(4.0*HU*HV)
      B4=1.0/(2.0*HU)
      B5=1.0/(2.0*HV)
      DO 50 K=2,N4
      CALL CBO (K)
      Z1=0.5*(RA(K)+RA(K+1))
      Z2=0.5*(ZA(K)+ZA(K+1))
      Z3=0.5*(RVA(K)+RVA(K+1))
      Z4=0.5*(ZVA(K)+ZVA(K+1))
      Z5=(RVA(K+1)-RVA(K))/HV
      Z6=(ZVA(K+1)-ZVA(K))/HV
      R20=Z3-Z1
      Z20=Z4-Z2
```

```
      DO 40 J=2,N1
C     COMPUTE DERIVATIVES OF VACUUM POTENTIALS
      XX1=PP(1,J+1,K)-PP(1,J,K)
      XX2=PT(1,J+1,K)-PT(1,J,K)
      YY1=PP(1,J+1,K+1)-PP(1,J,K+1)
      YY2=PT(1,J+1,K+1)-PT(1,J,K+1)
      FUU=B1*(C11*(XX1*XX1+YY1*YY1)+C33*(XX1*XX2+YY1*YY2)+C22*(XX2*XX2+Y
     1Y2*YY2))
      YY1=PP(1,J,K+1)-PP(1,J,K)
      YY2=PP(1,J+1,K+1)-PP(1,J+1,K)
      YY3=PT(1,J,K+1)-PT(1,J,K)
      YY4=PT(1,J+1,K+1)-PT(1,J+1,K)
      FVV=B2*(C11*(YY1*YY1+YY2*YY2)+C33*(YY1*YY3+YY2*YY4)+C22*(YY3*YY3+Y
     1Y4*YY4))
      YY1=PP(1,J+1,K)+PP(1,J+1,K+1)-PP(1,J,K)-PP(1,J,K+1)
      YY2=PT(1,J+1,K)+PT(1,J+1,K+1)-PT(1,J,K)-PT(1,J,K+1)
      YY3=PP(1,J,K+1)+PP(1,J+1,K+1)-PP(1,J,K)-PP(1,J+1,K)
      YY4=PT(1,J,K+1)+PT(1,J+1,K+1)-PT(1,J,K)-PT(1,J+1,K)
      FUV=B3*(C11*YY1*YY3+C12*(YY1*YY4+YY2*YY3)+C22*YY2*YY4)
      AKK=1.0+EP*RB1(J)
      AKL=AKK*ZLE
      ARU=RBU1(J)
      AZU=ZBU1(J)
      ARV=RBV1(J)
      AZV=ZBV1(J)
      EDE=(ARU*AZV-ARV*AZU)/AKL
C     COMPUTE PSI DERIVATIVES
      YY1=AL(NI,J+1,K+1)-AL(NI,J,K+1)
      YY2=AL(NI,J+1,K)-AL(NI,J,K)
      PUU=B1*(YY1*YY1+YY2*YY2)
      YY1=AL(NI,J,K+1)-AL(NI,J,K)
      YY2=AL(NI,J+1,K+1)-AL(NI,J+1,K)
      PVV=B2*(YY1*YY1+YY2*YY2)
      YY1=AL(NI,J+1,K)+AL(NI,J+1,K+1)-AL(NI,J,K)-AL(NI,J,K+1)
      YY2=AL(NI,J,K+1)+AL(NI,J+1,K+1)-AL(NI,J,K)-AL(NI,J+1,K)
      PUV=B3*YY1*YY2
C     COMPUTE FREE BOUNDARY OPERATOR
      PAS1=R(J,K)-Z3
      PAS2=Z(J,K)-Z4
      PAS3=ARU*ARU+AZU*AZU
      PAS4=PAS3+EDE*EDE
      AKL2=AKL*AKL
      PAS5=AKL2+ARV*ARV+AZV*AZV
      PAS6=ARV*ARU+AZV*AZU
      PAS7=PAS1*ARU+PAS2*AZU
      PAS8=PAS1*ARV+PAS2*AZV
      PAS9=PAS2*ARV-PAS1*AZV
      PAS10=PAS1*AZU-PAS2*ARU
      HB1(J)=(RB1(J)-Z1)*ZBU1(J)-(ZB1(J)-Z2)*RBU1(J)
      DD=A7*HB1(J)*(AK(J,K)+AK(J+1,K)+AK(J,K+1)+AK(J+1,K+1))
      DD2=DD*DD
      BV2=(PAS3*FVV+PAS5*FUU-2.0*PAS6*FUV)/(PAS4*AKL2)
      BP2=(PAS3*PVV+PAS5*PUU-2.0*PAS6*PUV)/(DD2*AKL2)
      AK(J,K)=AKK*(0.5*(BP2-BV2)+PB)
C     COMPUTE LAX-WENDROFF COEFFICIENTS
      YY1=RU(J,K)
      YY2=ZU(J,K)
```

```
      YY3=RV(J,K)-Z5
      YY4=ZV(J,K)-Z6
      YY5=HB1(J)
      DFD=(R20*YY2-Z20*YY1+0.5*(X(J,K)+X(J+1,K)+X(J,K+1)+X(J+1,K+1))*(PA
     1S1*YY2-PAS2*YY1))/YY5
      DUD=(R20*PAS2-Z20*PAS1)/YY5
      YY5=(PVV/DD2-FVV/PAS4)/AKL2
      YY6=(PUU/DD2-FUU/PAS4)/AKL2
      YY7=(PUV/DD2-FUV/PAS4)/AKL2
      YY8=BV2/PAS4
      FU=-BP2*DUD+YY8*(PAS7-(PAS9*EDE)/AKL)+YY5*PAS7-YY7*PAS8
      FV=(-YY8*PAS10*EDE)/AKL+YY6*PAS8-YY7*PAS7
      PAS7=ARU*YY1+AZU*YY2
      PAS8=ARV*YY3+AZV*YY4+ZZ*AKK*EP*PAS1
      PAS9=ARU*YY3+AZU*YY4+ARV*YY1+AZV*YY2
      EF=-EP*PAS1*EDE/AKK-(YY3*AZU-YY4*ARU+ARV*YY2-AZV*YY1)/AKL
      AKF(J,K)=EP*PAS1*(PB-0.5*(BP2-BV2))+AKK*(-BP2*DFD+YY8*(PAS7+EDE*EF
     1)+YY5*PAS7+YY6*PAS8-YY7*PAS9)
      AKFU(J,K)=AKK*FU
      AKFV(J,K)=AKK*FV
   40 CONTINUE
C     PERIODICITY CONDITIONS
      AK(1,K)=AK(N1,K)
      AKFU(1,K)=AKFU(N1,K)
      AKFV(1,K)=AKFV(N1,K)
      AKF(1,K)=AKF(N1,K)
   50 CONTINUE
      DO 60 J=1,N1
      AK(J,1)=AK(J,N4)
      AKF(J,1)=AKF(J,N4)
      AKFU(J,1)=AKFU(J,N4)
   60 AKFV(J,1)=AKFV(J,N4)
C     COMPUTE LAX-WENDROFF ITERATION
      DO 90 K=2,N4
      SUM1=0.0
      SUM2=0.0
      DO 70 J=1,N1
      RBU2(J)=RBU1(J)
   70 ZBU2(J)=ZBU1(J)
      CALL CBO (K)
      DO 80 J=2,N1
      X1=0.25*(AK(J,K)+AK(J-1,K)+AK(J,K-1)+AK(J-1,K-1))
      X2=0.25*(AKFU(J,K)+AKFU(J-1,K)+AKFU(J,K-1)+AKFU(J-1,K-1))
      X3=0.25*(AKFV(J,K)+AKFV(J-1,K)+AKFV(J,K-1)+AKFV(J-1,K-1))
      X4=B4*(AK(J,K)+AK(J,K-1)-AK(J-1,K)-AK(J-1,K-1))
      X5=B5*(AK(J,K)+AK(J-1,K)-AK(J,K-1)-AK(J-1,K-1))
      X6=0.25*(AKF(J,K)+AKF(J-1,K)+AKF(J,K-1)+AKF(J-1,K-1))
      X7=X1+0.5*DTB*(X1*X6+X2*X4+X3*X5)
      AA=ABS(X7)
      Y2=0.25*(RBU1(J)+RBU1(J-1)+RBU2(J)+RBU2(J-1))
      Y3=0.25*(ZBU1(J)+ZBU1(J-1)+ZBU2(J)+ZBU2(J-1))
      SUM1=SUM1+X7*Y3*(1.0-X(J,K))
      SUM2=SUM2-X7*Y2*(1.0-X(J,K))
      ERBO=AMAX1(AA,ERBO)
   80 X(J,K)=X(J,K)+DTB*X7
      SUM1=SUM1*HU
      SUM2=SUM2*HU
```

```
          AA=ABS(SUM1)
          BB=ABS(SUM2)
          ERBJ1=AMAX1(ERBO1,AA,BB)
C         COMPUTE VACUUM COORDINATE AXIS ITERATION
          RVA(K)=RVA(K)+DTB*SUM1*FAXV
          ZVA(K)=ZVA(K)+DTB*SUM2*FAXV
          X(1,K)=X(N1,K)
       90 X(N2,K)=X(2,K)
          DO 100 J=1,N2
          X(J,1)=X(J,N4)
      100 X(J,N5)=X(J,2)
          RVA(1)=RVA(N4)
          RVA(N5)=RVA(2)
          ZVA(1)=ZVA(N4)
          ZVA(N5)=ZVA(2)
          RETJRN
          END



          SUBROUTINE ASIN
C         EVALUATES FLUX AND MASS FUNCTIONS AND INITIALIZES AXIALLY
C         SYMMETRIC SOLUTION. THIS SUBROUTINE SHOULD BE CHANGED IF MORE
C         GENERAL PRESSURE AND ROTATIONAL TRANSFORM PROFILES OR AXIALLY
C         SYMMETRIC WALL SHAPE ARE DESIRED
          COMMON /AUX/ R2,Z2,RA,ZA,X(30),R(10,30),AL(10,30),PA(10,30),D(10,3
         10),E(10,30),F(10,30),G(10,30),R1(10,30),Z1(10,30),GR(10,30),GZ(10,
         230),DD(10,30),QR(10,30),QZ(10,30),A(30),B(30),C(30),H(30),RR(30),Z
         3Z(30),RU(30),ZU(30),RB(30),ZB(30),RBU(30),ZBU(30),XA(30),XC(30),XP
         4D(30),XPK(30),AK2(30),AKF(30),AKFU(30),GAXR,GAXZ
          COMMON /INP/ Q(10),QT(10),AM(10),PR(10),QO(10),PC(10),TC(10),BPP(1
         10),BTP(10),BET(10)
          COMMON /CPL/ NI,NJ,NK,EP,ZLE,GAM,SM,N1,N2,N3,N4,N5,NVAC,HS,HU,HV,P
         1I,RX,RY,E4,A1,A2,A3,A4,A5,A6,HU4,HV4,HUV,IC,IO,SA1,SA2,SA3,SE1,SE2
         2,SE3,DT,RA1,RA2,RA3,RE1,RE2,RE3,ENER,FAXIS,DG1,DG2,DG3,DH1,DH2,DH3
         3,NPLOT,I1
          COMMON /CVA/ NIV,FV1,FV2,NV,NP,OM,PM,HR,H1,H2,H3,C1,C2,N6,EVAC,ETO
         1T,A11,A22,A12,HR4,DTB,FAXV,OM1,OM2,PM1,PM2
          COMMON /FUNC/ ALF,RBOU,DELO,DEL1,DEL2,DEL3,DEL10,DEL20,DEL30,DEL22
         1,DEL33,PO,XPR,AMUO,AMU1,AMU2,AMP,FUR(7),FUZ(7),FUX(7),NRUN
          DO 20 J=1,N2
          U=(J-2)*HU
          U1=U+0.5*HU
          UP1=2.0*PI*U1
          X1=COS(UP1)
          X2=SIN(UP1)
          X3=COS(2.0*UP1)
          X4=SIN(2.0*UP1)
          X5=COS(3.0*UP1)
          X6=SIN(3.0*UP1)
C         COMPUTE WALL SHAPE
          RAO=1.0+DEL10*X1+DEL20*X3+DEL30*X5
          RADU=-2.0*PI*(DEL10*X2+2.0*DEL20*X4+3.0*DEL30*X6)
          RR(J)=RAD*X1
          ZZ(J)=RAD*X2
```

```
      RU(J)=RADU*X1-2.0*PI*X2*RAD
      ZU(J)=RADU*X2+2.0*PI*X1*RAD
C     COMPUTE FREE BOUNDARY FUNCTION
      IF (NVAC.LT.0) GO TO 10
      X(J)=RBOU
      GO TO 20
   10 X(J)=1.0
   20 CONTINUE
      RA=0.0
      ZA=0.0
      R2=0.0
      Z2=0.0
      GUM=1.0/GAM
      FQ=PI*RBOU*RBOU
      AA=AMAX1(ABS(AMUO),ABS(AMU1),ABS(AMU2))
      DO 50 I=1,NI
      S=(I-1)*HS
      X1=S*(1.0+ALF*S)/(1.0+ALF)
      X2=SQRT(X1)
C     COMPUTE ROTATIONAL TRANSFORM PER UNIT LENGTH
      AMU=(AMUO+AMU1*X2+AMU2*X1)/ZLE
C     COMPUTE FLUX FUNCTIONS
      IF (AA.GT.0.1) GO TO 30
C     IF AMU IS SMALL, TOROIDAL FLUX FUNCTION IS CHOSEN TO SATISFY
C     0.5*BT*BT+P=CONST
      PRES=PO*((1.-X1)**XPR)
      QT(I)=FQ*SQRT(1.0-2.0*PRES)*(1.0+ALF*2.0*S)/(1.0+ALF)
      GO TO 40
C     IF AMU IS LARGE, TOROIDAL FLUX FUNCTION IS CHOSEN TO SATISFY BT=1
   30 QT(I)=FQ*(1.0+ALF*2.0*S)/(1.0+ALF)
C     POLOIDAL FLUX FUNCTION IS CHOSEN TO SATISFY Q/QT=AMU
   40 Q(I)=AMU*QT(I)
      QQ(I)=Q(I)*Q(I)
C     COMPUTE INITIAL VALUES OF R AND PSI
      DO 50 J=1,N2
      R(I,J)=X2
      U=(J-2)*HU
   50 AL(I,J)=-U*QT(I)
C     COMPUTE MASS FUNCTION FROM GIVEN INITIAL PRESSURE DISTRIBUTION
      DO 60 I=1,N3
      S=(I-0.5)*HS
      X1=S*(1.0+ALF*S)/(1.0+ALF)
      X2=FQ*(1.0+2.0*ALF*S)/(1.0+ALF)
      PRES=PO*((1.-X1)**XPR)
   60 AM(I)=PRES**GUM*X2
      IF (NVAC.LT.0) GO TO 80
C     COMPUTE INITIAL VALUES OF VACUUM POTENTIAL
      DO 70 J=1,N2
      U=(J-2)*HU
      DO 70 I=1,NIV
   70 PA(I,J)=U
   80 CONTINUE
      RETURN
      END
```

```
      SUBROUTINE SURF
C     COMPUTES OUTER WALL SHAPE AND INITIALIZES 3-D SOLUTION
C     THIS SUBROUTINE SHOULD BE CHANGED IF MORE GENERAL 3-D WALL SHAPE
C     IS DESIRED
      COMMON RO(10,30,30),AL(10,30,30),XO(10,30,30),XL(10,30,30),R(30,30
     1),Z(30,30),RU(30,30),ZU(30,30),RV(30,30),ZV(30,30),X(30,30),RA(30)
     2,ZA(30),RN(30),ZN(30),RB1(30),RB2(30),ZB1(30),ZB2(30),RBU1(30),RBU
     32(30),ZBU1(30),ZBU2(30),RBV1(30),RBV2(30),ZBV1(30),ZBV2(30),HB1(30
     4),HB2(30)
      COMMON /AUX/ RR2,ZZ2,RRA,ZZA,XX(30),E1(10,30),E2(10,30),F1(10,30),
     1F2(10,30),G1(10,30),G2(10,30),P1(10,30),P2(10,30),Q1(10,30),Q2(10,
     230),V1(10,30),V2(10,30),U1(10,30),U2(10,30),UV1(10,30),UV2(10,30),
     3D(10,30),DD(10,30),BK(10,30),CA1(30),CA2(30),CB1(30),CB2(30),CC1(3
     40),CC2(30),CD1,CD2,CE1(30),CE2(30),CF1(30),CF2(30),CG1(30),CG2(30)
     5,CL1(30),CL2(30),CM1(30),CM2(30),CN1(30),CN2(30),XA1(30),XA2(30),X
     6B1(30),XB2(30),XC1(30),XC2(30),XG1(30),XG2(30),XN1(30),XN2(30),XD1
     7(30),XD2(30),XF1(30),XF2(30),XP1(30),XP2(30),XQ1(30),XQ2(30),SPR(1
     80)
      COMMON /POT/ RVA(30),ZVA(30),BPV(10),BTV(10),PT(10,30,30),PP(10,30
     1,30)
      COMMON /INP/ Q(10),QT(10),AM(10),PR(10),QQ(10),PC(10),TC(10),BPP(1
     10),BTP(10),BET(10)
      COMMON /CPL/ NI,NJ,NK,EP,ZLE,GAM,SM,N1,N2,N3,N4,N5,NVAC,HS,HU,HV,P
     1I,RX,RY,E4,A1,A2,A3,A4,A5,A6,HU4,HV4,HUV,IC,IO,SA1,SA2,SA3,SE1,SE2
     2,SE3,DT,RA1,RA2,RA3,RE1,RE2,RE3,ENER,FAXIS,DG1,DG2,DG3,DH1,DH2,DH3
     3,NPLOT,I1
      COMMON /CVA/ NIV,FV1,FV2,NV,NP,OM,PM,HR,H1,H2,H3,C1,C2,N6,EVAC,ETO
     1T,A11,A22,A12,HR4,DTB,FAXV,OM1,OM2,PM1,PM2
      COMMON /FUNC/ ALF,RBOU,DELO,DEL1,DEL2,DEL3,DEL10,DEL20,DEL30,DEL22
     1,DEL33,PO,XPR,AMUO,AMU1,AMU2,AMP,FUR(7),FUZ(7),FUX(7),NRUN
      PI2=2.0*PI
      DO 10 I=1,N3
   10 AM(I)=AM(I)*ZLE
      DO 20 I=1,NI
   20 Q(I)=Q(I)*ZLE
      DO 70 K=1,N5
      V=(K-2)*HV
      VV=2.0*PI*V
      X2=SIN(VV)
      X3=COS(VV)
      X4=SIN(2.0*VV)
      X5=COS(2.0*VV)
      X6=SIN(3.0*VV)
      X7=COS(3.0*VV)
C     COMPUTE INITIAL VALUES OF MAGNETIC AND VACUUM COORDINATE AXES
      RDI=FUR(1)+FUR(2)*X2+FUR(3)*X3+FUR(4)*X4+FUR(5)*X5+FUR(6)*X6+FUR(7
     1)*X7
      ZDI=FUZ(1)+FUZ(2)*X2+FUZ(3)*X3+FUZ(4)*X4+FUZ(5)*X5+FUZ(6)*X6+FUZ(7
     1)*X7
      RA(K)=RRA+AMP*RDI
      ZA(K)=ZZA+AMP*ZDI
      RVA(K)=RR2+AMP*RDI
      ZVA(K)=ZZ2+AMP*ZDI
      VP1=(V+0.5*HV)*PI2
      VP2=2.0*VP1
      VP3=3.0*VP1
      X1=COS(VP1)
```

```
      X2=SIN(VP1)
      DO 70 J=1,N2
      U=(J-2)*HU
      UP1=(U+0.5*HU)*PI2
      UP2=2.0*UP1
      UP3=3.0*UP1
      Y1=COS(UP1)
      Y2=SIN(UP1)
      Y3=SIN(UP2-VP1)
      Y4=SIN(UP3-VP1)
      Y5=SIN(UP3-VP3)
      Y6=SIN(UP2-VP2)
      COMPUTE WALL SHAPE AND DERIVATIVES
      RAD=1.-DELO*X1+DEL10*Y1+DEL20*COS(UP2)+DEL30*COS(UP3)-DEL3*COS(UP3
     1-VP1)+DEL22*COS(UP2-VP2)+DEL33*COS(UP3-VP3)
      RADU=-PI2*(DEL10*Y2+2.*DEL20*SIN(UP2)+3.*DEL30*SIN(UP3)-3.*DEL3*Y4
     1+2.0*0EL22*Y6+3.*DEL33*Y5)
      RADV=PI2*(DELO*X2-DEL3*Y4+2.*DEL22*Y6+3.*DEL33*Y5)
      R(J,K)=RAD*Y1+DEL1*X1-DEL2*COS(UP1-VP1)
      Z(J,K)=RAD*Y2+DEL1*X2+DEL2*SIN(UP1-VP1)
      RU(J,K)=RADU*Y1-PI2*RAD*Y2+PI2*DEL2*SIN(UP1-VP1)
      ZU(J,K)=RADU*Y2+PI2*RAD*Y1+PI2*DEL2*COS(UP1-VP1)
      RV(J,K)=RADV*Y1-PI2*DEL1*X2-PI2*DEL2*SIN(UP1-VP1)
      ZV(J,K)=RADV*Y2+PI2*DEL1*X1-PI2*DEL2*COS(UP1-VP1)
      COMPUTE INITIAL VALUES FOR R AND FREE BOUNDARY FUNCTION FROM
      AXIALLY SYMMETRIC SOLUTION PLUS PERTURBATION
      UP1=PI2*U
      UP2=2.0*UP1
      UP3=3.0*UP1
      XDI=FUX(1)*COS(VV)+FUX(2)*COS(UP2)+FUX(3)*COS(UP3)+FUX(4)*COS(UP2-
     1VV)+FUX(5)*COS(UP3-VV)+FUX(6)*COS(UP2-2.0*VV)+FUX(7)*COS(UP3-3.0*V
     2V)
      IF (NVAC.LT.0) GO TO 40
      X(J,K)=XX(J)*(1.0+AMP*XDI)
      DO 30 I=1,NI
   30 RO(I,J,K)=E1(I,J)
      GO TO 60
   40 X(J,K)=1.0
      DO 50 I=1,NI
      S=(I-1)*HS
   50 RO(I,J,K)=E1(I,J)*(1.0+AMP*(1.0-S)*XDI)
   60 CONTINUE
      COMPUTE INITIAL VALUES FOR PSI
      DO 70 I=1,NI
      S=(I-1)*HS
      AL(I,J,K)=E2(I,J)+Q(I)*V
   70 CONTINUE
      IF (NVAC.LT.0) GO TO 90
      COMPUTE INITIAL VALUES FOR VACUUM POTENTIALS
      DO 80 I=1,NIV
      DO 80 J=1,N2
      DO 80 K=1,N5
      PT(I,J,K)=(K-2)*HV
   80 PP(I,J,K)=F1(I,J)
   90 CONTINUE
      RETURN
      END
```

INDEX

Acceleration 17, 18, 19
—formula 29
—scheme 42
Admissible functions 33
Alfven transit time 20
Analytical solution 57
Artificial time 17, 19, 23, 29
Artificial viscosity 2, 33
Aspect ratio 59
Average 24, 30
Axial symmetry 18, 56, 67
—deformations 38
—program 41, 115
—solution 121

Beta 2, 58
Bifurcation 18, 52
Bilinear functions 25
Boundary condition 9, 15
Boundary quantities 94

Calcomp plotter 37
Cards 37
CDC 6600 37, 41
Characteristics 7
Coils 2, 59
Compatibility conditions 9, 15, 22
Compression ratio 18, 54, 59, 63
Computer 35
Computer code 27, 33, 36
Computing time 37
Conformal mapping 14
Conjugate gradient method 17
Conservation form 16, 22, 26
Constraints 4, 5
—,flux 3, 11, 59
—,mass 7
—,ergodic 7, 11

Containment 67
—,time 2, 63
Convergence factor 20
Convergence rate 19, 34, 57
Coordinate system 8
—,cylindrical 15
—,Lagrangian 9
Courant–Friedrichs–Lewy criterion 19, 28, 31, 41
Cross sections 50, 59, 63
Current 5, 44
—density 67
—,force-free 65
—,net 12
—,skin 49
—,toroidal 66
Cut 6
Cylinder 9, 38, 49

Data 37, 86
Density 4
Descent coefficients 29, 32, 41, 83
Destabilizing 33
Determinant 22
Diagonally dominant 27
Difference approximations 23
Difference equations 14, 22, 26
Diffuse plasma 49
Dimensionless radius 9
Dimensions 41
Direct methods 14
Dirichlet's integral 15, 25, 34
Dirichlet's principle 13
Discrete variational principle 22, 33
Discretization 22
Dispersion relation 19, 28
Dissipation 30, 42

Eigenfunction 18, 20
Eigenvalues 33, 52
Elliptic equations 11, 33
Energy 4
—inequality 32, 41
—,internal 7
—,minimum 17
—,potential 11
Equilibrium 1, 32
—,bifurcated 3
—solutions 4
—,stable 18
—,unstable 18, 50
Ergodic constraint 7, 11
Errors 37, 41
—,least-squares 19
—,roundoff 39
—,truncation 24, 33
Euler equations 5, 9, 10, 23
Evolution 32
Exact solution 49, 56
Existence 22, 58
Experiment 63, 69

Fast Fourier transform 34
Finite element method 15, 22, 25, 33
First-order accurate 23
First-order partial differential equation
    30
First variation 10, 17, 35
Flux 13
—constraints 3
—distribution 11
—function 6, 56
—poloidal 5, 43
—surfaces 9, 26, 40
—,toroidal 5, 43
Force balance 32
Force-free field 12
Fortran IV 37
Fourier analysis 18, 36, 80
Fourier coefficients 32, 40, 78
Free boundary 9, 14, 23, 40
—condition 17
—equation 125, 134
—radius 37
—variation 16

Free surface 5, 20, 31
FTN compiler 37

Garching 2, 3, 58, 63
Gauss–Seidel method 34
Gaussian distribution 59
Growth rates 19, 32, 44, 52, 64

Hadamard's variational formula 14, 31
Heat equation 34
Helical excursion 21, 40, 59
Helically symmetric 38, 63
—windings 3
Hölder's inequality 7
Hyperbolic 17, 31, 34

Inductance matrix 13, 27
Initial values 138
Input data 36, 78
Instabilities 63
—,exponential 51
—,mathematical 31
—,numerical 36
—,physical 31, 36
Integration by parts 23
Interpolation 115
Isar Tl-B 49, 58
Islands 6
Iterations 42, 126
Iterative scheme 23, 28, 32

Jacobian 7, 24, 28, 41
Jameson's scheme 35

Laplace's equation 15, 25, 33, 35
Larmor radius 2, 33, 63
Lawrence Berkeley Laboratory 65
Lax–Wendroff method 30, 125
Leap frog 30
Least-squares 106
Local minimum 4
Los Alamos Scientific Laboratory 2, 58

Magnetic axis 5, 9, 22, 27, 40
Magnetic field 9
Magnetic lines 6
Magnetohydrodynamics 4, 18
Magnetostatics 5, 10
Mappings 9
Mass 5, 36

Mesh 30, 39
—points 24
—,rectangular 25, 27
—refinement 33
—size 29, 37, 52, 63
Minimax 29, 32, 43
Mode 3, 20, 36, 40, 51
—,internal 54
—number 44, 66

Natural boundary condition 13
Neumann problem 15, 22
Nodal values 22
Nonlinear 2, 33, 51, 60
—saturation 3
Nonstandard type 11, 17, 33
Numerical computation 22

Operator 10, 17, 19, 28
—,elliptic 35
—,space 119
Orthogonal basis 13
Output 44, 107

Parameters 36
Periodicity 9
—conditions 16, 41
Periodic section 37
Perturbations 10, 14, 39
Plasma displacements 32
Plasma equations 88, 95
Plots 37, 44, 98
Poloidal fluxes 5, 39
Potential 12, 17, 35
Pressure 4, 38
Pressure profile 2, 49
Pressureless plasma 49, 59, 63, 67
Print 43, 98
—growth rates 112

Radius 37
Reciprocal variational problems 12
Relaxation factor 20, 29, 34, 43
Residuals 32
Resistivity 67
Restoring force 59
Rezoning 21
Richardson method 17, 29
Rotational transform 6, 23, 38
Run 37, 72

Saddle point 18
Saturation 18, 52
Scaling 23, 28
Screw pinch 56
Scyllac 2, 49, 58
Second-order accurate 23, 57
Sharp boundary 5, 49, 59, 65
Simulation 63
Spline 113
Stability 2
—,linearized analysis 18
—,numerical 31, 42
—theory 54
Stabilizing effect 64, 67
Stationary points 4
Steepest descent 2, 17, 19
Stellarator 3, 38, 49, 59
Storage 41
—core 37
Streak plots 60
Successive over-relaxation 20, 29, 34

Taylor series 30
Theta pinch 1
Three-dimensional 3
—calculation 37, 41
—equilibrium 33
—geometry 2, 12
—problem 6, 29
Time step 32, 41
Tokamak 1, 39, 49
Topology 6, 12, 54
Toroidal 5
—angle 8, 38
—directions 5
—equilibrium 2
—flux surfaces 6
—fluxes 5, 23, 39
—geometry 4, 9, 22
—nested surfaces 12
Torus 8, 37
Trajectories 63
Transformation 8
Triangular cross sections 3, 63

Uniqueness 9, 22
Unstable 18
—equilibrium 50, 60
—modes 39

Vacuum 27
—equations 122, 131
—region 11, 42
Validation 58
Variation of the independent variables 9
Variational principle 1, 4

Wall 139
—,conducting 5
—,outer 38, 59
—radius 37
—,stabilization 3, 67
Wave numbers 3
Weak solutions 12, 22, 33
Well-posed problem 4, 7

Lecture Notes in Economics and Mathematical Systems

Managing Editors: M. Beckmann, H. P. Künzi

Volume 66
A Theory of Supercritical Wing Sections, with Computer Programs and Examples

F. BAUER, P. GARABEDIAN, and D. KORN, Courant Institute of Mathematics and Computing, New York University, 251 Mercer Street, New York, NY, USA

1972. v, 211 pages, 19 illustrations. paper

"Mathematical methods for the design of supercritical wings, which depend on the numerical solution of the partial differential equations of two-dimensional gas dynamics, are developed. The main contribution is a computer program for the design of shockless transonic airfoils using the hodograph transformation and analytic continuation into the complex domain. The mathematical theory is described, and a manual for users of the programs is provided. Numerical examples are given and computational results are discussed, and the computer programs themselves are listed. The analysis routine can be used to ascertain whether the profiles behave well at off-design conditions, or to smooth coordinates and obtain a desirable shape more quickly when perfectly shockless flow is not essential."

International Aerospace Abstracts

Volume 108
Supercritical Wing Sections: A Handbook

F. BAUER, P. GARABEDIAN, D. KORN, and A. JAMESON, Courant Institute of Mathematics and Computing, New York University, 251 Mercer Street, New York, NY, USA

1975. v, 296 pages, 84 illustrations. paper

This handbook is a sequel to *A Theory of Supercritical Wing Sections, with Computer Programs and Examples.*

From the Preface: ". . . Since the completion of the first volume . . . some effort has been made to improve our airfoil design program. A number of more desirable airfoils have been designed. In addition several of our wing sections have been tested in wind tunnels. We should like to make this material available here, since it is more convenient to use the design program in conjunction with data for a fairly broad range of examples. Moreover, we have developed new analysis programs that supersede our previous work."

Volume 150
Supercritical Wing Sections III

F. BAUER, P. GARABEDIAN, and D. KORN, Courant Institute of Mathematics and Computing, New York University, 251 Mercer Street, New York, NY, USA

1977. v, 179 pages. paper

The third volume surveys computational flow research on the design and analysis of supercritical wing sections.

Springer-Verlag   New York   Heidelberg   Berlin

Springer Series in Computational Physics

Editors: W. Beiglböck, H. Cabannes, E. Orszag

Numerical Methods in Fluid Dynamics

M. HOLT, University of California, Berkeley, CA, USA

1977. viii, 253 pages, 107 illustrations, 2 tables. cloth

Information: Numerical problems in gas dynamics are treated in the first part of this monograph. The discussion of finite difference methods is concentrated on hyperbolic systems. The author describes the present status of two approaches developed in the USSR, both based on the method of characteristics: the method of Godunov and the BVLR method of Rusanov and his associates. Also included are techniques of Butler and Sauer. In subsequent chapters, the author describes the methods of integral relations introduced by Dorodnitsyn, Telenin's method, and the method of Lines—techniques based on polynomial or series representations to the unknowns—all applied to problems in fluid dynamics. Numerous applications and samples of numerical solutions of model problems are presented.

Contents: General Introduction. The Godunov Schemes. The BVLR Method. The Method of Characteristics for Three-Dimensional Problems in Gas Dynamics. The Method of Integral Relations. Telenin's Method and the Method of Lines.

Springer-Verlag   Berlin   Heidelberg   New York